# DINOSAUR PLOTS

## AND OTHER INTRIGUES IN NATURAL HISTORY

# DINOSAUR PLOTS

## AND OTHER INTRIGUES IN NATURAL HISTORY

LEONARD KRISHTALKA

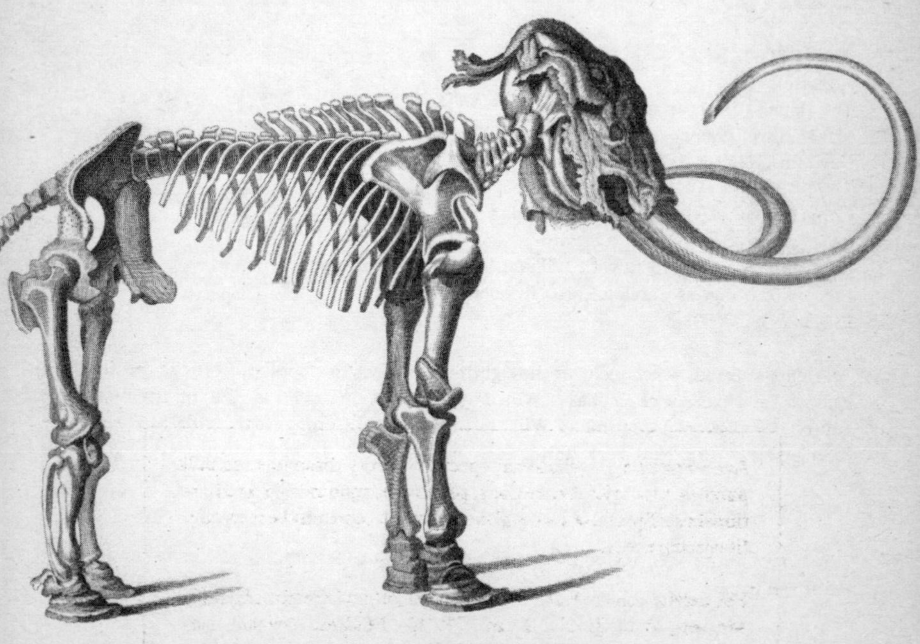

AVON BOOKS NEW YORK

AVON BOOKS
A division of
The Hearst Corporation
105 Madison Avenue
New York, New York 10016


Cover illustration by Joseph Devito
Published by arrangement with William Morrow and Company, Inc.
Library of Congress Catalog Card Number: 88-13622
ISBN: 0-380-70998-8



First Avon Books Trade Printing: April 1990

AVON TRADEMARK REG. U.S. PAT. OFF. AND IN OTHER COUNTRIES, MARCA REGISTRADA, HECHO EN U.S.A.

Printed in the U.S.A.

OPM 10 9 8 7 6 5 4 3 2 1

For Beth

XXIII.—Ideal landscape of the Eocene Period.

*A paleontological intrigue from primeval Europe, 50 million years ago, featuring a herd of browsing three-toed, snouted palaeotheres, extinct descendants of the earliest horses. (From Louis Figuier,* The World Before the Deluge, *D. Appleton and Co., New York, 1867)*

# Preface

Paleontology hurtles through time like no other scientific machine. It starts in the Archean, three and one half billion years ago, with petrified algae. It plunges forward through towering, stacked, geologic layers, each flashing a glimpse of life on earth during an extinct moment. The array of past species is staggering, the mass of their remnants even more so: pollen, leaves, wood, roots, seeds, tests, carapaces, teeth, bones, burrows, impressions, shells, skin, feathers, footprints.

Paleontology is the intrigues left by a vanished world imperfectly preserved. It is a scattershot of skeletons and sediments and ancient ecological plots; it is tales of origins, extinctions and tumultuous descent. There are long geologic bounds of evolutionary stillness broken by staccato bursts of sudden revolutionary change; there is the imperceptible pace of a slow, steady, serial procession of life.

Paleontology spirals from arcane minutiae to universal principle. The arcane is the cusp pattern on the molar teeth of a 70-million-year-old fossil mammal from Montana called *Purgatorius*, the earliest known primate and perhaps the progenitor of us all. The principle, or one of them, is how those cusps, that primate and all of life came to be.

Paleontology also fashions its own culture and characters, from bonehunters zigzagging, often stumbling, over that scattershot of clues to brilliant discoveries and incredible boners, ingenious science and disingenuous charlatans.

In 1982, I began a natural-history column called "Missing Links" in *Carnegie Magazine*, which is Carnegie Institute's lean, bimonthly answer to *Smithsonian.* The first article was about pigs and their indirect contribution to evolutionary thought, everything from French truffles and symbiosis to the age of the hominid-bearing strata in East Africa. I wondered then about the column's long-term evolutionary odds: Would there be another article for the next issue, and yet another, and from what corner of the mind would they emerge, because at that moment all the corners seemed frighteningly barren—what paleontologists call unfossiliferous.

There is no shortage of intrigues in natural history and, luckily, the articles emerged. I say luckily although I haven't polled my readers for their opinion. And they are opinionated. One reader wrote:

> I have read your page "Art on the Rocks." I think that your French words used by art critics are particularly enlightening. I have always envisioned you as a doddering bibliophile from some Eurasian country or perhaps India. Only recently have I come across a photograph of you which makes you appear as a college sophomore.

Inexplicably, a number of others also imagined me to be some donnish geezer from a country in East Gondwanaland. I answered this reader and set the record straight.

> The truth about my appearance is somewhere in between your images of me. I was once a college sophomore (and looked

like it) and one day I will, no doubt, be a doddering bibliophile. Currently, I'm a middle-aged bibliophile in desperate need of a haircut. My only connection with Eurasians and India is that Krishtalka usually appears in the phone book between Krishnaswami and Krisiewicz. My broken knowledge of French comes from growing up in Montreal, Canada.

This book is a collection of selected articles from the "Missing Links" column, each of which has been substantially updated and revised. The major players in these articles are early man, mammoths, cows, pigs, dinosaurs and other hordes of bygone beasts. They link bonehunters, badland rocks, hoaxes, taxonomy, art, the stars, the Red Queen, the Montreal Expos, the Montreal Canadiens, Leakey, Linnaeus, Darwin, evolution and time. There is a smattering of Greek and Latin, courtesy of the animal names, and French, courtesy of art. I apologize for the French. In Montreal they taught us that *hors d'oeuvres* is a sandwich cut up into a hundred pieces.

My wife, Beth Krishtalka, was my companion for these ideas in Pittsburgh, and in the field in Wyoming and Kenya. She was the first to audition them in conversation and on paper and she, more than anyone else, helped puncture the snooty critical platitude that the popular cannot be serious. For special words, whimsy and history, I reached for my brother, Aaron Krishtalka.

Maria Guarnaschelli, my superb editor at William Morrow, suggested the book, saw me through it, and adapted her wonderful literary eye to these essays. In science, as in art, the beauty is in the detail, and here she ensured that language and attitude matched thought. *Carnegie Magazine* editor, R. Jay Gangewere, did likewise.

My colleagues at The Carnegie Museum cheerfully passed their professional eyes over these articles: Mary Dawson, Richard Stucky, Elizabeth Hill, Doug Swarts,

Andrew Redline and Craig Black (now on the Pacific Plate, at the Los Angeles County Museum of Natural History). Elizabeth Hill unearthed a wealth of paleontological archival material concerning The Carnegie Museum, Carnegie's Dinosaurs, Dinosaur National Monument and other historical matters. Mary Ann Schmidt's assistance was invaluable.

# Contents

CONTENTS

# CHAPTER 1

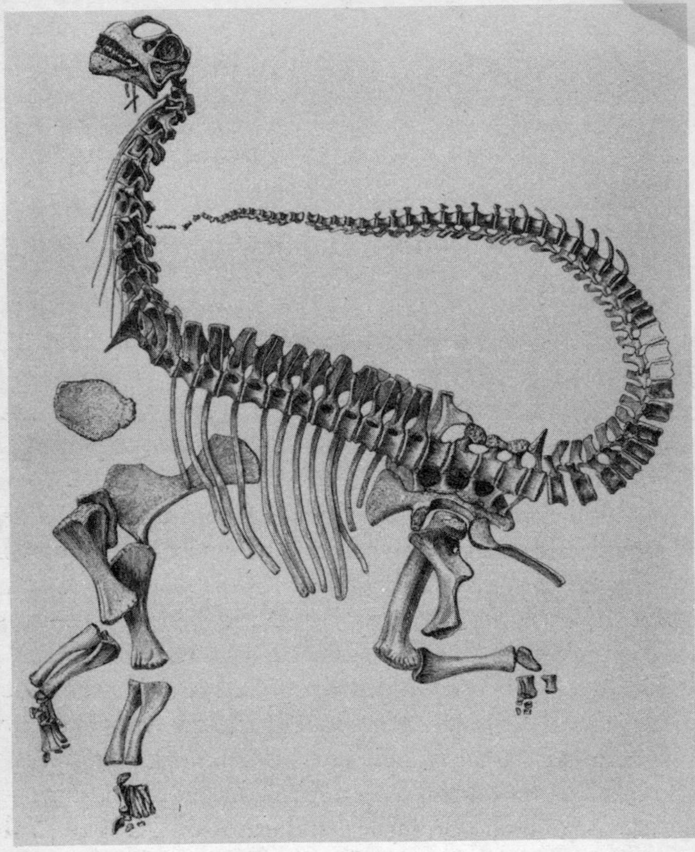

*This seventeen-foot skeleton of a juvenile* Camarasaurus lentus *is the finest specimen of a sauropod dinosaur ever found. The young* Camarasaurus *died 135 million years ago and was buried and preserved in a sandbar of an ancient river in what are now the badlands of Dinosaur National Monument, Utah, a site discovered in 1909 by Carnegie Museum paleontologist Earl Douglass. (From Charles W. Gilmore,* Memoirs of the Carnegie Museum, *10, 1925)*

# Dinosaur Plots

It was not so long ago that dinosaurs were dead and cold. Museums, books and monster movies dressed them in industrial gray or fatigue green. Slowly the ancient beasts lumbered across the landscape, the cloddish, plodding sluggards of yesteryear. Dinosaurs set the cultural standard for the overgrown, the passé, the evolutionary dead end. Again and again they played the extinct straight-beast to the latest technological razzle-dazzle. Dinosaurs were the metaphor for the dumb and defunct: Their bodies were immense, but their brains could fit in a walnut.

In the late sixties, though, the revolution arrived. Leading the paleontological renaissance of dinosaurs was Robert T. Bakker, then a student at Yale University, who challenged the traditional notion of dinosaurs as dim-witted, cold-blooded, ponderous behemoths. Bakker, Yale's John Ostrom and other students of dinosaurs forced paleobiologists over the next decade to reevaluate the bones and flesh of dinosaur dogma: their behavior, posture, locomotion, brain size, rate of growth, bone structure, metabolism, physiology and ecology.

What emerged was a new manifesto for the Mesozoic, the 150-million-year Age of Reptiles. Dinosaurs were hot.

New reconstructions dressed them in the latest dazzling colors: *Tyrannosaurus* and *Triceratops* looked as if they had just dashed through an Earl Scheib paint job, flashing Day-Glo red and silver lightning bolts. Steaming articles in the scientific and popular press promoted a fiery image for these ancient reptiles. The huge plant eaters (sauropods: *Brontosaurus*, *Diplodocus* and their ilk) were frisky and limber, strutting with tails off the ground and long necks towering high in the air. They grew quickly to adulthood and roamed in thundering herds, protecting the young at the center. The two-legged meat eaters (carnosaurs: *Tyrannosaurus*, *Allosaurus* and their kin) and their smaller, bird-footed cousins (the coelurosaurs) were fast, ferocious, intelligent predators, able to outrun and outsmart their herbivorous prey. Some species (*Deinonychus*) may have hunted in packs, ganging up on hapless victims, such as their unarmed herbivorous cousins, and ripping them apart with huge, sickle-shaped foot-claws. Other species (*Oviraptor*) plundered nests for the eggs of horned dinosaurs. The plates on plated dinosaurs (stegosaurs: *Stegosaurus* and others) were solar panels that imparted "warm-bloodedness." The dome on dome-headed dinosaurs (pachycephalosaurs: *Stegoceras* and kin) became battering rams in butting contests over territory and mates. The crests on crested dinosaurs (hadrosaurs: *Corythosaurus*, *Lambeosaurus* and others) flashed and honked sexual notes of passion and power across the marsh flats; mounds of fossilized eggs, hatchlings and infants of one hadrosaur (*Maiasaura*) indicate the practice of parental care in nesting colonies. Adult horned dinosaurs (ceratopsians: *Triceratops* and others) were only a few steps shy of the Indianapolis 500: They could charge and gore a predator at 40 mph.

The new saga of dinosaurs was exhilarating. It had an epic sweep of 130 million years of geologic time, a spectacular cast of dinosaurian monsters from Montana to Mongolia, and a plot resplendent in reptilian blood and guts. After more than 200 years of collecting and studying dinosaurs, the revolution had produced a "heretical" portrait[1] of dinosaurian life, loves, locomotion, sex, social structure, feeding, fighting, brains, brawn, body language and behavior. It also produced dinosaur "piggy" banks, cups, saucers and cookie cutters, dinosaur chocolates, planters, aprons and hot pads, dinosaur sweatshirts, T-shirts and caps, dinosaur towels, washcloths, shower curtains, bed sheets and pillowcases, dinosaur fireplace ornaments, dinosaur Christmas-tree ornaments, dinosaur greeting cards, calendars, pen and pencil sets, mobiles and kites, blow-up dinosaurs, windup dinosaurs, motorized dinosaurs and dinoburgers.

The revolution had a final, stunning coup de grâce: birds. Dinosaurs, the proverbial evolutionary dead end, had sired a feathered legacy. From among the early, bipedal, bird-footed coelurosaurs had arisen *Archaeopteryx*, the earliest known bird, 120 million years ago. Research on dinosaurian anatomy by John Ostrom and others revealed that *Archaeopteryx* was a feathered dinosaur: All that distinguished it from its reptilian predecessors was its feathers (which came from the same embryonic tissue as scales) and its fused collarbones (the modern "wishbone"). Indeed, one of the five known specimens of *Archaeopteryx* had lain in a museum drawer in Eichstatt, Switzerland, for twenty-three years—misidentified as the small dinosaur *Compsognathus*, its evolutionary cousin.[2]

Every revolution has its short-haired, wait-and-see skeptics. For many paleontologists, including Peter Dodson[3] at the University of Pennsylvania and Alan Charig[4] at the British Museum of Natural History in London, the hot-blooded conclusions about dinosaurs needed to be more discriminating. Not all dinosaurs were warm-blooded. Not all measures of warm-bloodedness in dinosaurs, such as predator/prey ratio and bone histology, are foolproof barometers of metabolic rate. The available evidence suggests that only some of the dinosaurs, notably the hunting coelurosaurs and carnosaurs, may have been as active and hot-blooded as mammals. The dome-headed, horned and crested dinosaurs, with moderate-sized brains and a grinding dentition, were at least "lukewarm-blooded," that is, they had higher metabolic rates and activity levels than is typical of living reptiles. Other dinosaurs, certainly the jumbo sauropods (*Diplodocus, Brontosaurus*), were more likely cold-blooded physiologically but, because of their tremendous bulk, could have been passive heat sinks in the semitropical regimes of the Mesozoic. The essence of the hot- versus cold-blooded debate is the diversity of dinosaur types. Dinosaurs were too diverse in shape, size and structure to afford generalizations about their physiology.

In the past few years, the dinosaur stampede has returned, this time to a different tune—the death march of sudden, catastrophic extinction. The disappearance of the dinosaurs has been called "the grandest whodunit in natural history."[5] Most of the facts in the case of the dinosaurs' demise are known. About 65 million years ago dinosaurs ceased to exist on earth. The extinction episode marks the end of the Cretaceous, the last geologic period of the Mesozoic Era (the Age of Reptiles), and the beginning of the Tertiary, the first period of the Cenozoic (the Age of Mammals). The roster of death at this 65-million-year-old boundary reads like a who's who of the Cretaceous animal kingdom: on land, all dinosaurs and pterosaurs (flying reptiles); in the sea, many of the one-celled, free-floating organisms called plankton, the ammonites (a group of one-shelled molluscs), many clams, sea lilies, starfish and moss animals, and all plesiosaurs, mosasaurs and ichthyosaurs (marine paddle-finned and fishlike reptiles). All perished before or at the end of the Cretaceous Period. Some investigators estimate that three fourths of all animal and plant species disappeared during a geologic instant 65 million years ago. A sudden extinction of such magnitude is catastrophic—and spawns catastrophic explanations.

But on the other side of the ledger, a host of organisms were not affected by the scourge of extinctions: land dwellers, such as birds, crocodiles, lizards, turtles, snakes and amphibians, eluded death. So did most plants and pouched mammals (marsupials) and almost all non-pouched mammals (placentals). Also untouched were freshwater animals such as insects, shellfish and bony fishes as well as fishes in the ocean.

A few years ago, Dale Russell, a paleontologist at Canada's National Museum of Natural History in Ottawa, pinned the extinctions on a lethal blast of heat and gamma rays from the deadly explosion of a supernova close to the earth 65 million years ago. This scenario was the latest in a long line of global Cretaceous disaster stories vying for scientific acceptance and aimed at solving the whodunit. As with previous cataclysmic tales, this one soon became extinct for lack of evidence. But shortly thereafter, a new candidate for catastrophe surfaced: the asteroid-impact theory of dinosaur extinction.

The asteroid thesis goes like this. About 65 million years ago a colossal asteroid, about nine miles in diameter, penetrated the atmosphere and crashed into the earth. The impact created a huge crater, shattered the asteroid and threw the resultant tonnage of asteroid-earth dust into the stratosphere. The dust cloud encircled the earth, blotted out the sun and interrupted the process of photosynthesis in terrestrial plants and marine plankton. All but the most resistant plants perished, initiating a chain reaction of extinction in the ecosystem—a biological domino effect. On land, first the plant-eating dinosaurs and then their predators, the carnivorous dinosaurs, died out in the face of a disappearing food supply. A similar disaster occurred in the seas, starting with the death of the photosynthetic plankton and moving up the marine food chain to many invertebrate animals and the marine reptiles. After a few months the dust settled out of the atmosphere and sunlight once again reached the earth's surface. The surviving plants and phytoplankton could photosynthesize again, the remaining vertebrates and invertebrates could feed, and life proceeded on an earth now devoid of dinosaurs and the other doomed organisms. Is this ingenious science or voodoo paleobiology?

The first clues that inspired the asteroid-earth collision theory were circumstantial. Scientists from the University of California, Berkeley, while measuring the concentration of certain heavy elements in deep-sea sediments in Italy, discovered that 65-million-year-old marine rocks are composed of an unexpected clay layer sandwiched between two limestone layers. The limestones, or chalky (calcareous) sediments, contain a mixture of rock and the chalky remains of dead plankton—skeleton and shells that are continually sinking to the ocean floor. The limestone layers represent either side of the Cretaceous-Tertiary boundary (a term geologists shorten to "K-T boundary"). The clay layer, on the other hand, is devoid of calcareous skeletons and marks the actual K-T boundary, the time of the alleged mass extinctions. As is typical, the limestones showed only a trace amount (about 0.3 parts per billion) of a particular dense, metallic element called iridium. Iridium is extremely rare in the earth's crust because it settled to the core when the earth was still in a molten state. But, to the astonishment of the Berkeley group, the thin clay bed registered 30 times as much iridium as the limestones. Tests of similar rock sequences (limestone-clay-limestone) spanning the K-T boundary in Denmark and New Zealand showed 160- and 20-fold increases in iridium levels in the boundary clay.

The boundary-clay layer is a double mystery. For one, its mere presence speaks of a sudden interruption in the normal, day-to-day deposition of chalky limestone on the ocean floor near the time of the extinctions. For another, its lode of iridium either came from deep within the bowels of the earth through some unknown process or had an extraterrestrial origin: Asteroids are rich in iridium and other rare, heavy metallic elements.

For Luis Alvarez, the Berkeley Nobel Prize–winning physicist, and his colleagues the choice was clear: The K-T clay layer was born in space. In a 1980 article in *Science*, they claimed that 65 million years ago a monstrous asteroid smashed into the earth, disintegrated on impact and sent a massive cloud of intermingled earth and asteroid dust into the atmosphere. After a short time, perhaps only a few months, the dust fell to earth, onto land and into the sea, as a discrete geologic stratum of clay well endowed with the asteroid's iridium. Furthermore, the timing of this event with the disappearance of dinosaurs and other organisms could not be sheer coincidence—the dust-cloud blackout had caused the mass extinctions at the end of the Cretaceous.[6]

Since 1980, this "catastrophist" school of paleobiologists has amassed more support for its extraterrestrial theory: a worldwide record of iridium anomalies spiking 65-million-year-old rocks laid down on land and in the sea; computer-modeled effects of stratospheric dust clouds;

global "nuclear winter" scenarios and their deadly biological consequences; and evidence of highly pulverized ("shocked") quartz grains in the boundary clay, the result of the tremendous force and heat of the asteroid impact.

A few paleontologists have joined the asteroid school, but most have not, inviting charges that earthbound scientists prefer earthbound explanations—a natural prejudice toward extraterrestrial theories. Well, let me countercharge that some catastrophists are intellectual ballerinas in loose tights and some catastrophe theories are small ideas injected with growth hormone. In science, as in art, the beauty is in the detail, and the detail in this instance is unkind to an asteroid catastrophe.

One fundamental problem with the asteroid scenario is scope: Too few organisms became extinct. A second is timing: Dinosaurs, according to the fossil record, went out with a whimper, not a bang. And third: The iridium isn't necessarily celestial.

Problem one. A catastrophe of the magnitude of an asteroid impact should have exterminated most groups of organisms, not just a preferred few. Yet the extinctions struck organisms of selective stripes, implying selective causes. Specifically, the dust-cloud blackout and sudden onset of a nuclear winter would have plunged temperatures on land to about 0 degrees Fahrenheit or lower, bringing snow and ice to what had been tropical forests. If this happened, why did the extinctions bypass most groups of marine life—snails, crustaceans, bony fish, sharks, skates, rays—as well as fishes and invertebrates living in lakes and streams? On land, how did most of the mammals, birds, amphibians, and reptilian kin of the dinosaurs—crocodiles, lizards, snakes and turtles—escape extinction? It's difficult to explain their survival in the face of destroyed marine and terrestrial food chains, chaotic ecological upheavals and a runaway nuclear winter.[7]

Another effect of the dust cloud would have been acid rain and snow from the production of large amounts of nitrogen oxides. This too should have decimated the freshwater organisms, but the fossil record shows this did not occur. Worse, the nitrogen oxides should have temporarily erased the earth's ozone layer, exposed terrestrial organisms to lethal doses of ultraviolet light and caused virtually instant, mass death. Yet most land plants and animals breathed into the Age of Mammals. In short, the discrepancies between the predicted and the observed are too great; the asteroid as angel of death may be a sham rock.

Problem two. We may be putting the cart before the dinosaur. Where is it written that dinosaurs disappeared in a sudden cataclysm of mass extinctions? Not in the rocks. The fossil record implies the contrary, that hundreds of dinosaur species came and went without fanfare throughout their 130-million-year reign on earth; that during the last 10 million years of the Mesozoic (from about 75 to 65 million years ago), the number of dinosaur species dwindled from thirty to twelve or fewer.[8] Twelve species is not mass extinction. In fact, the last individual dinosaurs appear to have died *before* the time of the iridium anomalies and before the asteroid hit, if it hit. Their bones are found in rocks that underlie, and thus predate, the iridium-rich 65-million-year-old stratum.

Nor does the nondinosaur fossil evidence spell sudden catastrophe. Among sea animals, shellfish (ammonites, clams and others) as well as marine and flying reptiles gradually dwindled in numbers and diversity throughout the last part of the Cretaceous and were almost—if not completely—extinct before the alleged asteroid impact. Mammalian communities show gradual evolution across the 65-million-year K-T boundary, as do many species of plants. Some paleobotanists maintain that fewer than 50 percent of late Cretaceous plants became extinct, and of these, many were already on the decline or had disappeared well before the end of the Cretaceous. Land plants do not show a global pattern of extinction. Rather, plants at higher latitudes had a higher extinction rate than those in the tropics, which were least affected. The reverse should have occurred in a nuclear winter or some other global catastrophe. Temperate floras, unlike tropical ones, are more resistant to sudden environmental changes as a result of such adaptations as long-term seed dormancy and underground tubers. The only large-scale catastrophic extinctions at the 65-million-year boundary involve the one-celled marine plankton and selected invertebrates along the marine food chain: lamp shells (brachiopods), sea lilies, starfish and moss animals.

The catastrophists counter that the fossil record is biased by a declining number of good localities in late Cretaceous rocks. For that reason, they say, there are fewer dinosaur fossils and species known from 65-million-year-old rocks than there are from ones 75 or 80 million years old. Accordingly, the picture of gradual decline in diversity and abundance of specific fossil reptiles, invertebrates and plants is artificial and not a reflection of real events. But this kind of reasoning begs the question: If the fossil record is too poor to establish gradual extinction, why is it then a trusted measure of catastrophic extinction? No doubt, the fossil record of late Cretaceous organisms needs to be improved and refined to get a clearer picture of the timing, magnitude and pattern of animal and plant extinctions. At this point, however, the fossils say the pattern was mosaic, the magnitude large but selective and the timing gradual for all but a few groups of organisms.

Problem three of the impact theory involves guilt by association: iridium and asteroids. An asteroid may have hit the earth 65 million years ago, but the geological jury is still out on this one.[9] The iridium anomalies at the K-T boundary could not, as Alvarez claims, have been laid down in a few weeks following one impact; they occur throughout a thickness of sediments that represents about 100,000 to 1 million years of time. Furthermore, the boundary clays at different localities have different chemical compositions, which is inconsistent if the source of the boundary clay was one global dust cloud. Finally, the concentration of copper, zinc, arsenic, antimony, selenium and other elements in the boundary clay is too high for an asteroid but correct for a terrestrial origin.

Geologists don't need to reach into space for iridium. They found an earthly source for iridium in some of the explosive eruptions of Hawaii's Kilauea volcano, which produced volcanic dust ten thousand times as iridium-rich as other eruptions. Apparently, some volcanic explosions tap an iridium source in the deep mantle of the earth, whereas other lava eruptions do not.[10,11]

The shocked-quartz particles so characteristic of the boundary clay were once thought to be unimpeachable evidence of the extreme force and pressure that only an asteroid impact could generate. But according to data from the Mount St. Helens eruption, explosive volcanoes can also fracture and splinter the crystal structure of quartz grains. As confirmation, shocked quartz has since been identified in volcanic-ash beds in Sumatra and California.

In any case, if a colossal asteroid smashed into the earth's surface 65 million years ago, where is the crater? Computer simulations estimate the size of the impact crater at 90 to 120 miles across, which ought to be the eighth wonder of the world. Craters fill the geologic record, but none fit the K-T date nor these gigantic dimensions. It's not likely a huge crater on land or the continental shelf would be covered over to the point of being undetectable. The impact buffs answer that the asteroid probably crashed into the ocean floor where, subsequently, the crater was buried under millions of years of sediment. The flaw here is that ocean-floor sediment is virtually quartz-free, so wherefore the shocked quartz?

The upshot is that the upheavals 65 million years ago were earthbound. Rocks that date to the final millennia of the Cretaceous reveal two spectacular and probably interrelated phenomena: first, tremendous episodes of mountain building, with uplift and deformation of the earth's crust; second, a broad retreat of the world's oceans and, consequently, a precipitous drop in sea levels. Vast areas of continental coastline were exposed, destroying coastal reefs, continental shelves and shallow inland seas—the habitats of the marine organisms that ultimately vanished. On land, the tropical regimes gave gradual way to temporary harsher, seasonal climates, abetted by the chilling atmospheric effects of volcanic-ash clouds. Some plants slowly succumbed to extinction, as did the last of the dinosaur species. During the last 10,000 years of the Cretaceous, exceptional and successive volcanic eruptions left a signature of clay, iridium and shocked-quartz grains across the 65-million-year-old rocks. Disastrous climatic effects ensued, which wiped out the more vulnerable land plants and many of the oceanic plankton and their ecological dependents: shellfish, lamp shells and sea stars.

This is the best we can do for now. As in all whodunits, the culprit (or culprits) of the extinctions left some incriminating clues, but the trail is 65 million years cold and many of the geologic fingerprints are smudged. One bonus of the asteroid hoopla was the discovery of iridium anomalies, first in 65-million-year-old rocks and later in many other geologic strata. It sent geologists down new avenues of research strewn with iridium-spiked rocks and earthbound theories for how the iridium got there.

Meanwhile, many of the hot-blooded dinosaurian heresies have become orthodoxy, except in most museums where new exhibits are prohibitively expensive and take almost as long to evolve as the beasts themselves. I brought the latest writings on dinosaurs to The Carnegie Museum and read them amid the players in Dinosaur Hall—their skeletons are the measure of hot-blooded theories and cold-blooded death. *Newsweek*[12] was so enraptured with catastrophic dinodeath that it pushed up the R.I.P. date from 65 million to 26 million years. Other articles were replete with simple notions of extinctions: death stars, planet X and 26-million-year cycles of organismal doom which not only killed the intended victims but also cleaned the Augean stables. I am wary of simple theories. They may have appeal, but one difference between nature and its interpreters is that nature is not simple.

# Chapter 2

*The shrew,* Sorex, *might have been named* Katharina *had Linnaeus borrowed from Shakespeare instead of Latin. (From William Jardine,* The Naturalist's Library. Mammalia, *Vol. XVII:* British Quadrupeds, *Chatto & Windus, London, 1875)*

# The Naming of the Shrew

In 1904, an entomologist named Kirkaldy published an innocent-looking paper in a British journal that had all the appearances of a yawner. In it he assigned new scientific names to a swarm of different insects in the Order Hemiptera, a group that includes the bugs leaf hoppers, plant hoppers and stainers. But, amid the monotonous litany of Latin and Greek names appeared *Peggychisme* (pronounced "Peggy-kiss-me"), for one of the stainers, and *Polychisme*, for another. One kind of plant hopper became *Florichisme*, two bedbugs *Dolychisme* and *Ochisme*, and three seed bugs *Marichisme*, *Elachisme* and *Nanichisme*. For the leaf hoppers, plant bugs and assassin bugs, Kirkaldy changed his sexual preference, naming *Alchisme*, *Zanchisme* and *Isachisme*. The paper, or rather the frivolous names, bent the starched collars of British entomology. Kirkaldy obviously had more on his mind than the serious taxonomy of insects, although it's curious that he avoided "*Pleaschisme*" and "*Dontchisme*." Had the plays been written, Kirkaldy would certainly have named "*Chismekate*" and "*Chismet.*" In 1912, the Zoological Society of London officially condemned his choice of scientific names.

Spats over plant and animal names have been with us

ever since early man realized he was just part of the organic scenery and began naming and classifying his world of living and inert objects. Trouble is, a babel of names arose from two sorts of confusion: The same species of plant or animal was referred to by a different name in different areas; or different species were lumped under the same name. Today, for example, a crayfish in some quarters is a crawdad, crawcrab, crawdabber, crawdaddy, crawjinny, crawldaddy, crawpappy and craydad in others. A gopher, on the other hand, can connote any one of a dozen different rodents, a turtle and other critters.

So, early natural philosophers, beginning with Aristotle in the fourth century B.C., set themselves the task of describing and classifying all known plants and animals according to a precise, formal system of names. Many classifiers were driven by other motives, such as codifying what they perceived as the Creator's handiwork, His roster of organic achievements before Man. Whatever the motive, these naturalists discovered an immense, unimagined diversity of living and fossil organisms. Aristotle recognized 550 kinds of animals, and Renaissance herbals list between 250 and 600 kinds of plants.[1] By the mid-1700s those numbers had jumped to 6,000 for plants and 4,000 for animals in the two major classifications of the time: Carolus Linnaeus's *Species plantarum* (published in Stockholm in 1753) and *Systema naturae* (tenth edition, 1758).

Linnaeus, often credited with being the father of taxonomy (the science of classification), culminated the centuries-long passion for naming and cataloging the "essential kinds of life" on earth, from organisms encountered in the neighboring meadow to those that explorers were bringing back from Asia, Africa and new worlds across the oceans. The sheer number of plant and animal species led Linnaeus and his contemporaries to abandon the biblical notion of a Creator's roster of essential kinds and to realize that they were classifying much more than "as many species as were created in the beginning." Such maturation of thought has so far eluded modern creationists.

Plaudits aside, Linnaeus apparently wasn't up on his Shakespeare. Had he lent an ear to the stage as well as to nature Linnaeus might have taken a cue from Shakespeare's play and named the genus of shrew "Katharina" instead of *Sorex*, the Latin word. Ever since Linnaeus, naturalists have blown one shrewd opportunity after another. Newly discovered kinds of shrews were given scientific names like *Blarina*, *Crocidura* and *Domnina* but not Katharina. When we finally got around to using "Katharina" as a scientific name for an animal, we bungled the beast altogether. In 1847, J. E. Gray bestowed *Katharina* on a new type of chiton, a shelled mollusc. The situation was salvaged somewhat in 1928 when an entomologist named Girault honored the bard and his stinging prose by naming a particular type of parasitic wasp *Shakespearia*.

Linnaeus's work wrought order from natural chaos, setting forth a system of names that today is still scrupulously followed by botanists and zoologists. His scheme was brilliant yet simple. All names were in Latin and every species of organism received a two-part name, or binomial. The first name (called the genus, or generic name) denoted a grouping of creatures of an essentially similar "kind"; the second name represented the individual or specific "kind." Both names together described the species, the basic unit of living organisms. Thus, the genus *Sorex* stood for shrews of a certain ilk, namely, those that shared many basic physical similarities. Within *Sorex*, Linnaeus defined three distinct species: *Sorex araneus* (from Europe), *Sorex cristatus* (from Pennsylvania), and *Sorex aquaticus* (from eastern North America).[2] Similarly, Linnaeus placed all animals of the ox "kind" in the genus *Bos* (Latin for "ox") and within it distinguished five species.

Man was no exception to the Linnaean system. Linnaeus formally classified "our kind" as the genus *Homo* (Latin for "man"), with one known species, *Homo sapiens* ("man the wise"). Since 1758, two other species of *Homo*, both extinct (but not necessarily less wise), have been established from fossil remains in Africa and Asia: *Homo habilis* and *Homo erectus*.

Ever since Linnaeus, botanists and zoologists have done a booming business in names, adding vast numbers of new genera and species of fossil and living organisms to his original classification. Most of the names are still rooted in Latin or Greek words, and botanists, ever faithful to Linnaeus, persist in describing new species of plants in Latin. But zoologists have long strayed from the Roman tongue. As early as 1873, an American paleontologist, Joseph Leidy, named a new fossil carnivore *Sinopa*, the Blackfeet Indian word for "small fox." Other naturalists borrowed the Indian name *Ondatra* for the muskrat, the Mongolian term *Ochotona* for the pika, and the native Siberian word for the extinct mammoth, *Mammuthus*.[3]

Two animal names taken from the Sioux language rival Russian for tongue twisting: *Ekgmowechashala* (pronounced "Ig-uh-moo-wee-cha-sha-la") means "monkey" or "little cat man," and is a fossil primate. *Sunkahetanka* (pronounced "Sung-ka-hee-tahn-ka"), a fossil, doglike carnivore, means "large-toothed dog." The Sioux names are intentional. The fossil bones of both animals were found in 26-million-year-old rocks in the Wounded Knee area of South Dakota, the site of a Sioux massacre ninety-five years ago.

For most people, scientific names are convincing proof

that biologists set out to create an incoherent system of arcane babble and succeeded. There is mild justification for this impression, given such polysyllabic Latin and Greek bongers as *Udabnopithecus*, *Baryphthengus*, *Sysphinctostoma* and the like. The evidence for a dimwit amid the taxonomists mounts when one reads that in 1926 a student of crustaceans (the group of shrimp, crabs and waterfleas) actually named a new species *Cancelloidokytodermogammarus* (*Loveninuskytodermogammarus*) *loveni*. Like Linnaeus, he probably thought that this moniker redounded to the greater glory of God. No doubt the Poles who named one of their towns Szczebrzeszyn thought likewise.

Actually, scientific names are much like Mahler's music: They're better than they sound. Behind the names lurk homages to people, geography, ecology, geologic age, mythology or a unique piece of anatomy. An Argentine paleontologist, Florentino Ameghino, had a penchant for naming new genera of extinct South American mammals after eminent biologists: *Asmithwoodwardia*, for A. Smith Woodward, a renowned British paleontologist;[4] *Josepholeidya*, for Joseph Leidy; *Thomashuxleya*, for Thomas Huxley, and so on. In 1932, one of the giants of evolutionary biology, George Gaylord Simpson, returned the favor and named a new South American fossil mammal *Florentinoameghinia*.

Thomas Jefferson, aside from being the third president of the United States, was an avid amateur paleontologist. His studies of fossil mammoths brought mammoth recognition: The last of the American Ice Age species of mammoth now carries the name *Mammuthus jeffersoni*. It's a good thing no Russian premier ever dabbled in paleontology. Otherwise we might be stuck with "*Ursus yurivladimirovichandropovus*" for some extinct species of Russian bear. As it is, we are stuck with *Khrushchevia,* a petrified fibrous organism found attached to a 500-million-year-old fossil coral from New Mexico. It reminded the describer, R. H. Flower, of a fossil wart, a compliment he bestowed in 1961 on the Russian premier of the time.

Insults in species names are rare, but irreverence is not. A few years ago University of Chicago paleontologist Leigh Van Valen labeled a new fossil doglike animal *Arfia.* Another paleontologist, Richard Estes, from San Diego State University, revealed his taste in Scotch by naming *Cuttysarkus,* a 70-million-year-old fossil lizard. Too bad the Scotch isn't as aged. Bonehunters with less breeding might have opted for "*Boonesfarmia,*" and gin drinkers for "*Beefeaterus.*"

Mary Dawson, a paleontologist at The Carnegie Museum of Natural History, opted for a mythological pun in describing a new species of the extinct carnivore *Daphoenus* (pronounced "Da-feen-us"). She must have jolted the art world and curators at the Louvre by naming the animal *Daphoenus demilo.* From mythology also comes *Pan* for the chimpanzee, *Diana* for a certain monkey and *Venus* for a clam. Leigh Van Valen, of *Arfia* fame, borrowed from the Icelandic Edda for *Ragnarok,* and from Tolkien's Silmarillion for *Thangorodrim,* two genera of extinct, primitive mammals.

Animal names are also like those of hockey players. For example, if you see a bunch of guys in the rink whose names end in ". . . off" or ". . . ov," you can also bet they skate fast, pass with precision, fly Aeroflot, wear jerseys adorned with hammer and sickle and play for Big Red. Animal names too, especially their suffixes, are often deliberate evolutionary and anatomical giveaways. Shrew names, for example, end in *sorex* ("shrew" in Latin), as in *Microsorex,* the "tiny shrew." Rodent names end in *mys* or *mus* ("mouse" in Latin or Greek). Rabbits end in *lagus* ("hare" in Greek), marsupials in *delphys* (Greek for "womb" or "pouch"), horses in *hippus* (Greek), dogs in *cyon* (Greek), turtles in *chelys* (Greek), snakes in *ophis* (Greek) or *aspis* (Latin), lizards and dinosaurs in *saurus* (Greek), and ape-like fossil man in *pithecus.* Many animals end in just plain *therium* (Greek for "wild beast"), as in *Megatherium,* a gigantic extinct sloth.

Some names combine geology or geography with these endings, as in: *Miohippus* and *Pliohippus,* fossil horses from the Miocene and Pliocene epochs; *Mytonolagus,* the oldest known North American fossil rabbit from a locality called Myton Pocket in Utah; and *Armintodelphys,* an extinct opossumlike marsupial from 50-million-year-old rocks near the town of Arminto, Wyoming (population 5).[5] Similarly, *Tsinglingomys* is a fossil rodent found in the Tsingling Mountains in China, *Alamosaurus* is a dinosaur from Texas, and *Utahia* is an extinct primate from Utah. Remains of the oldest known primate, *Purgatorius,* were recovered from 70-million-year-old rocks appropriately in the Hell Creek Formation, near Purgatory Hill, Montana.

Some names evoke the food habits and anatomy of the beasts. *Ornitholestes* (Greek for "bird robber") and *Oviraptor* (Latin for "egg plunderer") were small, bird-footed dinosaurs with an obvious penchant for poached eggs. *Carpolestes* was an extinct "fruit-robbing" (Greek) primate from North America, whereas *Merychippus* (pronounced "Merry-kippus") was a "grass-chewing" (Greek) fossil horse. At Christmas parties, sozzled paleontologists have been known to abandon all horse sense and launch into a rendition of "Have yourself a merry merychippus." The living platypus, *Ornithorhynchus,* has a Greek "duckbill" for a name, but one of the duck-billed dinosaurs, *Corythosaurus,* is named for the "helmetlike" (Greek) crest on its skull. Some extinct mammals have an enamel crest on their cheek teeth that resembles a *V*, or the Greek letter *lambda*, and are named accordingly: *Lambdotherium*, *Pantolambda*, *Sinolambda*. No one has yet had the temerity to name a creature "*Legolambda*." One of the winning name combinations must be *Cupidinimus nebraskensis*, a fossil rodent from the rocks of the Valentine Formation in Nebraska.

These names do more than promote Latin and Greek. They allow paleontologists and biologists to talk to one another without confusion about specific animals and plants. But Linnaeus's naming of species and genera was only act one in his ordering of nature. He also grouped similar genera into a higher category called an order. For example, the genera of man, apes and monkeys composed his order Primates (Latin for "first" or "chief" creatures). In turn, he combined similar orders into a class—Primates and seven other orders made up his class Mammalia. Similar classes were allocated to a kingdom, of which Linnaeus had three: animal, vegetable or mineral. He put fossils into the mineral kingdom, not realizing that they were also remains of past animals and plants. In short, all known organisms were classified from species to kingdom in this standard hierarchy, according to their degree of physical and anatomical resemblance to one another. It was Linnaeus's last act, and a smash hit in recognizing the patterns in nature.

A hundred years later, in the spotlight of evolutionary thought that followed Darwin, it became clear that nature had more to tell. Physical and anatomical similarities among creatures were of two main types. Some physical similarities between organisms were extensive and intricate, arising from a close evolutionary kinship—descent (and genetic inheritance) from a common ancestor. Other physical resemblances were more limited and superficial and not due to evolutionary genealogy; rather, they arose independently through chance or because the organisms evolved and live in similar habitats—the fishlike body shape and "fins" of sharks (fishes) and whales (mammals) are a good example.[6]

Accordingly, new classifications of plants and animals were developed and Linnaeus's work was rewritten with an evolutionary pen. Organisms were still grouped into species, genera, orders and classes, but only according to particular features—those physical and anatomical resemblances that advertised an evolutionary link. The names of these groups—Mammalia, Primates, Insecta, Reptilia—now became code words for millions of years of evolutionary lineages, family trees and particular anatomical traits.

The class Mammalia, for instance, is shorthand for all animals with hair, mammary glands, a four-chambered heart, two sets of teeth, a diaphragm, three middle-ear bones, and so on—a baggage of similarities inherited by all mammals by virtue of their common, evolutionary roots that reach back 200 million years. The Mammalia, in turn, contain some thirty-four orders of living and fossil mammals, each with its own name (e.g., Primates, Carnivora, Rodentia) and each a separate branch on the mammalian evolutionary tree.

One of those thirty-four branches, the order Insectivora, split off about 70 million years ago and sprouted the families of hedgehogs, moles, tenrecs, golden moles and shrews. The fossil record tells us that by the time Shakespeare started writing and Linnaeus started naming, over 100 species of shrews had already evolved and become extinct. Of the 250 or so living shrews, Linnaeus correctly recognized one, and Shakespeare, another. No matter if they named it *Sorex* or Katharina. A shrew by any other name might not smell any sweeter, but it would tell the same tale just as well.

# CHAPTER 3

*A Lecture.*—"You will at once perceive," continued PROFESSOR ICHTHYOSAURUS, "that the skull before us belonged to some of the lower order of animals; the teeth are very insignificant, the power of the jaws trifling, and altogether it seems wonderful how the creature could have procured food."

*A paleontological about-face: a school of student ichthyosaurs (ancient fishlike reptiles suddenly re-evolved from extinction) deciphering the skull and principles of a new fossil species,* Homo sapiens. *(From Francis T. Buckland,* Curiosities of Natural History, *Rudd & Carleton, New York, 1859)*

# Lesser-Known Principles

On October 8, 1909, a large earthquake shook the Kulpa Valley in Yugoslavia. Afterward, a Yugoslavian geophysicist named A. Mohorovičić examined the earthquake records, especially the speed of the seismic waves as they spread outward and downward from the epicenter. He noticed that suddenly, at a certain depth within the earth, the velocity of the seismic waves shot up from 3.6 miles per second to 4.8 miles per second. Earthquakes elsewhere gave the same results: When seismic waves reached a point about 20 miles below continents and 7 miles below ocean basins, their speed jumped more than a mile per second.

This discovery earned Mohorovičić immortality: The seismic speed-bump zone deep within the earth was dubbed the Mohorovicic discontinuity. It has also earned Mohorovičić the wrath of geology students, who curse the fates of science that allowed a major geological structure to be discovered by a Yugoslavian with a circumbendibus for a name. They can't do an end run around the blasted thing. It's too basic to bypass. The Mohorovicic discontinuity appears to be the dividing line between two of the earth's main layers, the lighter crust and the heavier, underlying mantle. The jump in velocity of the seismic wave as it crosses the discontinuity is due to a 33 percent increase in density of the mantle rock.

Geologists, following Orwell's dictum that whoever controls the past controls the future, finally shortened the Slavic moniker to "Moho" and the crust/mantle boundary to "Moho discontinuity." Obviously science, like totalitarian states, will revise the past to make the present more acceptable. But this kind of merciful simplification is rare. The scientific literature is choked with the polysyllabic jawbreakers of lesser-known principles named after their principals.

Science should adopt a basic principle against naming a new discovery—a Law, Rule, Discontinuity or whatever—after a person with more than, say, a seven-letter appellation. Slavic names, where vowels appear about as often as the Comet Kohoutek, should be stricken or pared to the essential. Moho from Mohorovičić is a good example, and one that should have been followed for the Comet Koho. A precedent here is the New York Giants football team of the late fifties and early sixties, whose behemoth defensive line of Dick Modzelewski, Andy Robustelli and Jim Katcavage was officially cut back to Mo, Ro and Ko. (Only New York sportswriters can divine a Ko from Katcavage.)

The scourge of the Yugoslavian confloption didn't end with Mohorovičić. In 1941, we were saddled with Milutin Milankovitch, a Yugoslavian geophysicist and astronomer, who formulated an important astronomical theory of climatic change to explain the series of Ice Age glaciations. Initial credit should go to James Croll, a mid-nineteenth-century Scottish naturalist, who early on attributed the Pleistocene pattern of glacial advances and retreats to changes in the earth's orbit, which affected the amount of solar radiation reaching the earth's surface.

Milankovitch elaborated Croll's ideas into mathematical

models based on three orbital features: (1) variations in the earth's orbit around the sun (called orbital eccentricity); (2) changes in the angle that the axis of the earth makes with the orbital path around the sun (termed the obliquity); and (3) the wobbling of the earth's axis of rotation (called the precession). He determined that all three variations are cyclical. On average, the orbital eccentricity goes from elliptical to almost circular every 95,000 years, the obliquity from 22 degrees to 24.5 degrees every 41,000 years (we're now at 23.5 degrees), and the precession cycles every 26,000 years. The interaction of the three cycles controls the seasonal amount of solar radiation that reaches the earth and causes major fluctuations in seasonal climates.

When Milankovitch extrapolated his results back through Pleistocene time, three periods emerged—185,000, 115,000 and 70,000 years ago—when the confluence of the cycles produced conditions most conducive to glaciation, namely, when the high latitudes received much lower levels of solar radiation during the summer months. His three dates are now known to coincide with periods of major ice growth; his cycles have been confirmed, refined and christened the Milankovitch cycles, which, like Moho and the three Giants, would be better off as the Mo cycles.

Geologists should thank the seismic gods for the Richter scale of measuring earthquakes. We could have been stuck with using the Rossi-Forel intensity scale, a measure of earthquake force devised in 1878 by the Italian M. S. de Rossi and the Swiss F. A. Forel, or worse, its 1931 successor, the Wood-Neumann modified Mercalli intensity scale. Luckily, Charles F. Richter, an American seismologist, devised a better, seven-letter measure of earthquake magnitude in 1935. The scale on the Richter scale runs from 1 to 9 and is logarithmic, so that an earthquake registering

6 is ten times more powerful than one measuring 5 and one hundred times more forceful than one measuring 4. Generally, earthquakes of 4.5 or more cause some damage, and those registering 7 or higher are severe. The devastating San Francisco quake of 1906 rang up a 7.8 on Richter's scale; the 1964 Alaska earthquake rumbled through at a whopping 8.4.

Many of the great scientists and their discoveries come in at seven letters (or fewer), like Newton's law, Boyle's law, Planck's constant, Halley's comet and the Doppler effect—even other discontinuities, like the Birch discontinuity, a deeper cousin (540 miles) of the Moho, or the Conrad discontinuity, which occurs at different depths in the earth where the speed of a seismic wave increases from 3.6 to 4 miles per second.

Some exceptions to the seven-letter rule are acceptable, as in Einstein's theories of relativity, where the extra letters are really relative, or in Bernoulli's Principle, which would be seven letters if it didn't have the extraneous *o* and *l*. In any event, it sounds tantalizing enough to describe the physics of cooking pasta, but it was named for Daniel Bernoulli (1700–1782), a Swiss mathematician, physicist and physician who discovered that as the speed of a moving fluid increases, the pressure within the fluid decreases. Bernoulli's Principle proved essential for the invention of the carburetor, the aspirator and political speeches. In aspirators, the moving fluid is water; in carburetors and politicians it's gas.

Another forgivable exception to the seven-letter fiat is Bergmann's rule, published in 1874 by a European zoologist of the same name. Its eight letters are more than compensated for by the five letters in its twin principle, Allen's rule. Bergmann's rule describes a geographic cline or gradient for warm-blooded animals, namely, birds and mammals. Within one species, individuals in colder environments tend, on the average, to have larger bodies; those in warmer climes are smaller. As a result species with a wide geographic distribution, like rodents and rabbits, show a gradient of increasing body size from the equator to the poles.

Allen's rule was coined in 1877 by Joel Asaph Allen, an American naturalist, and deals with the size of body extensions in warm-blooded animals. Individuals of a species in colder climates, or higher latitudes, tend to have shorter, smaller extremities (tails, ears, limbs, bills) in proportion to their body size than their mates in the tropics. Critics of Canada's Progressive Conservative government maintain that the cranial extremities of Cabinet members are consistent with Allen's rule.

Essentially, the Bergmann-Allen principles predict that from the equator to the north pole, mammalian species will have progressively larger bodies and shorter appendages. As freshman zoology students, our mnemonic rule for keeping these two rules straight was Bergmann for "body" and Allen for "appendages." Both rules rely on the ratio of an animal's surface area to its body volume. An elephant, for example, has a much smaller surface area/body volume ratio than a mouse. When an animal balloons in size, its internal bulk or body volume (a cubic dimension) quickly outstrips its surface area or area of skin covering (a square dimension). As traditionally interpreted, the Bergmann-Allen rules work because the skin covering of warm-blooded animals is the major heat exchanger with the outside. Natural selection among mammals and birds in cold climates would favor a proportionately smaller surface area (larger body, shorter extremity) to minimize heat loss; the need to radiate more heat in warmer locales would favor a relatively larger surface area (smaller body, longer extremity).

No rules are without exceptions—itself a rule that spawned the legal profession—and both Bergmann's and Allen's rules are no exception. Burrowing animals, such as moles and voles, are protected from the cold in their subterranean digs and consistently flout the body demands of Bergmann's rule. The tails and wings of birds, which are not involved in heat loss, disobey Allen's rule. Predator/prey relationships among warm-blooded species (the dynamics between the hunters and the hunted), as well as locomotion and food requirements, may outweigh heat regulation in governing the size of bodies and appendages. Also, the nostrils of some warm-blooded animals are more involved in heat regulation than is the skin. A final insult to Bergmann's warm-blooded rule is that a number of cold-blooded animals also "obey" it despite the minimal role their body size plays in temperature regulation.

Perhaps the most fundamental challenge to Bergmann's and Allen's rules is the radical notion that the physical environment has little to do with driving organic evolution. This challenge is called the "Red Queen," an eight-letter hypothesis as coldly pragmatic as its namesake and as elegant as its source, Lewis Carroll's *Through the Looking Glass.* It was formulated in 1973 by Leigh Van Valen, an eight-letter University of Chicago paleontologist. His extra letter was quickly pardoned when he gave paleontology *Arfia, Ragnarok, Purgatorius* and other wonderfully named fossil creatures.[1]

Van Valen's Red Queen draws on the evolutionary work of Darwin, Lyell, George Gaylord Simpson and Robert H. MacArthur. Its premise is simple: The evolution of species is essentially independent of changes in the physical environment. Rather, evolution is fueled by the interactions among the species themselves, especially through their competition for limited resources, such as food and space. For example, if one species achieves an evolutionary advantage in its dentition, which allows it to process a particular type of food more efficiently, all other ecologically similar species must keep up or lose out in the competition for that food resource. The same applies to other adaptations, such as speed of escape from predators or increase in body size. The Red Queen proclaims biological capitalism with a vengeance: The major component of the environment powering the evolution of species *is other species*, not changes in physical conditions, such as climate or topography.

In short, ecologically related species are on a natural treadmill of continuous evolutionary change driven by the species themselves—much like airline fare cuts and gas wars. Lewis Carroll's Red Queen said it best: "Now here, you see, it takes all the running you can do, to keep in the same place."

Integral to Van Valen's Red Queen is the "macarthur," his longish but solid unit that measures discrete episodes of evolution. It pays homage to the late Robert H. MacArthur, who in 1967, along with E. O. Wilson, produced the arid-sounding twenty-six-letter "Theory of Island Biogeography" to explain the dynamics and diversity of island faunas. The "Bikini Theory" would have been a catchier, more proper (less than seven-letter) name. Whatever the name, the theory helps clothe the Red Queen with evidence.

MacArthur and Wilson found that the number of species on an island depends on its surface area, the local extinction rate and the distance of that island from a source of immigrants, such as the mainland or other islands. After islands are invaded, settled and saturated by immigrant species, a dynamic equilibrium sets in: The total number of species on the island remains the same, but there is constant species turnover and change through extinction, immigration and competition. Essentially, the species keep running just to stay in place.

Van Valen coined the macarthur, appropriately, as a unit measure of this process. It denotes the rate at which discrete evolutionary phenomena, such as the extinction or origin of a species, occur with a 50 percent probability over five hundred years. For example, based on calculations for fossil and living primates, the genus of man, *Homo*, has a .00022 chance of becoming extinct in the next five hundred years.[2] Alas, the macarthur didn't catch on with paleobiologists as an evolutionary unit and is itself approaching extinction.

At the same time Van Valen was devising the Red Queen, a University of Arizona ecologist, Michael Rosenzweig, independently conceived the same idea and stamped it the "Rat Race."[3] In the rat race of evolutionary terminology, Red Queen has fared better than macarthur and outcompeted Rat Race.

It is apt, of course, that the name of a discovery mimic the name of the discoverer, and so the principles of Batesian mimicry and Mullerian mimicry are naturals. In 1862, Henry Walter Bates (1825–1892), an English naturalist and explorer, discovered that ordinary species of butterflies closely mimicked the coloration of species of unpalatable or poisonous butterflies and so deceived birds and other potential predators. This kind of evolutionary coattailing has made bees, wasps and hornets the most mimicked natural subjects. It is also tailor-made evidence for the Red Queen hypothesis. The credo of Batesian mimicry reads: Look unappetizing and you won't get eaten, a principle that has not yet spread to organic food.

Mullerian mimicry was named for Fritz Muller, a German naturalist, who, seventeen years after Bates, observed that different species of obnoxious, poisonous or venomous animals, mostly insects, come to mimic each other's color patterns. That way, members of an entire guild of prey species present the same warning pattern to predators and stand a better chance of not being the victim of even one taste test.

To be fair, many of the lesser-known principles conform to the seven-letter rule: Bragg's law (the reflected angle of X rays through a crystal depends on crystal planes and X-ray wavelength); Bouguer reduction (a gravity correction for altitude above sea level); Stephan's law (the total energy radiated per square centimeter of surface area of a glowing object is proportional to the fourth power of its temperature); Wien's law (the hotter an object, the shorter the wavelength of its peak emission), and others.

But there are many that don't. Perhaps the worst violators of the seven-letter rule are the Germanic flappers like "Hunter-Schreger Bande," which is a kind of Schmelzmuster (enamel arrangement) in the teeth of fossil and living mammals. Other winners are the Lotka-Volterra model, which describes the population dynamics among predators and their prey (it too is a Red Queen subject); Przewalski's horse, the wild horse of Mongolia and Xinjiang; the Kolmogorov-Smirnov test, which sounds like a vodka analyzer but is a statistical test for the distribution of natural populations; Fahrenholz's rule, which states that the evolutionary relationships of animals can be inferred from that of their parasites; the Hertzsprung-Russell diagram, a chart of the temperature and luminosity of stars; and Mitscherlich's formula, which calculates the amount of fertilizer for the highest agricultural yield. It's lucky that one of the most widely used statistical measures in natural history—"the bell-shaped curve"—didn't get labeled after its mentor, Aleksandr Mikhailovich Liapunov. By comparison, even Krishtalka's seven-letter rule is starting to sound good.

# CHAPTER 4

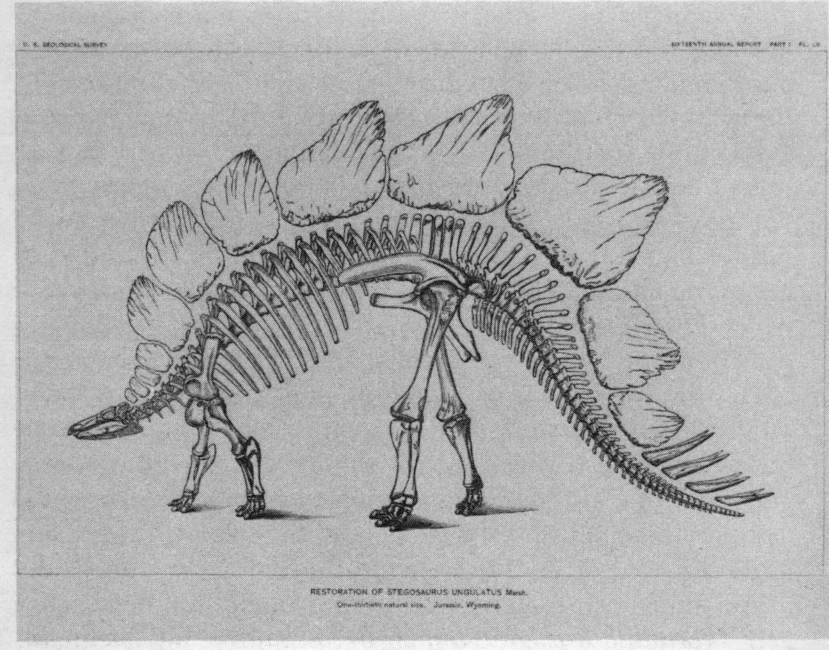


Stegosaurus ungulatus, *the famous plated dinosaur from the late Jurassic, 135 million years ago. A* Stegosaurus *running at 15 miles per hour hit with a force of 41,000 newtons. (From O. C. Marsh, "The Dinosaurs of North America," U. S. Geological Survey,* 16th Annual Report, *1895)*

# Begging the Question

Last year, in an attack of dementia, we asked children who visited Dinosaur Hall in The Carnegie Museum of Natural History to jot down their most pressing questions about dinosaurs, drop them in a box and return the following week for the answers, which we posted on a nearby bulletin board. The exercise proved two things: First, we were more accurate than the IRS, which in 1987 admitted to muffing one of every four questions at tax time; second, there were two sides to every question but no bottom.

Predictably, we were bombarded with requests for dinosaur vital statistics: Which was the biggest dinosaur? Which was the smallest? The tallest? The shortest? The smartest, the dumbest, the fiercest, the fattest, the fastest, the friendliest, the strongest, the longest, the slowest, the heaviest, the skinniest, the weakest, the lightest? How long ago did they live? How long did they live? How much did they weigh, eat, fight and sleep? How many bones are there in *Tyrannosaurus*, *Allosaurus*, *Brontosaurus* and *Stegosaurus*? Which dinosaur had the most teeth? And so on.

Even the IRS can provide dinosaur stats—their size, weight, physical appearance, cranial endowment and anatomy—by consulting David Norman's *The Illustrated*

*Encyclopedia of Dinosaurs* (New York: Crescent Books, 1985) and other works.[1] More challenging were the questions from a whopping 9 percent of the children, who were either precocious or fronting for embarrassed parents: they wanted the latest on dinosaur sex, and some of their questions crossed the line between science parlance and massage parlance.

**Did dinosaurs fall in love?**

I doubt it. Dinosaurs, despite their small brains, knew that love is an emotional island surrounded by expenses.

**Did dinosaurs make love with the opposite sex?**

Yup. It may be expensive, but much of evolution *is* sex with the opposite sex. It's the only way to explain a 130-million-year genealogy of *sauruses* as well as more than a billion-year history of sexual (eucaryotic) life on earth. Sexual reproduction breeds not only fun but also genetic reshuffling, resulting in physical differences among individuals and, ultimately, evolutionary change and the origin of new species. Some female lizards often forgo the fun and reproduce parthenogenetically, the animal version of the virgin birth. But among crocodiles and birds, the closest living evolutionary relatives of dinosaurs, the females like sex with the opposite sex and, by inference, so did the dinosaurs.

**How did the large dinosaurs reproduce?**

Traditional reconstructions of dinosaurs depict the huge sauropods—*Apatosaurus*, *Camarasaurus*, *Diplodocus*, *Brachiosaurus* and their kin—as sluggish, lumbering beasts, hardly candidates for sexual acrobatics. More recent studies by paleontologist Robert Bakker paint these behemoths as flashy, warm-blooded, agile and limber, much like Disney's dancing hippos in *Fantasia.*[2] Hot-blooded or not, male dinosaurs must have "mounted" female ones during copulation so that the eggs could be fertilized inside the female before being covered with the protective shell or leathery skin. One reconstruction by paleontologist L. B. Halstead features an X-rated picture of a male and female in a credible mating position for internal fertilization.[3]

The fertilized eggs were then either laid in brood nests or retained in the mother's body for gestation and hatching. Fossil egg nests of *Maiasaura,* a duck-billed dinosaur found in Montana, and *Protoceratops,* a horned dinosaur from Mongolia, attest to the former. Of the 350 or so known species of dinosaurs, only 7 plant eaters (representing duckbills, horned dinosaurs, primitive prosauropods and the huge sauropods) have been associated unequivocally with egg laying. As for the other 343 species, we have to be satisfied with the well-worn scientific cliché: Absence of evidence for egg laying is not evidence of absence of egg laying.

Direct evidence for the gestation and hatching of young inside the maternal dinosaur is not something that is preserved in the fossil record. The fact that some snakes and lizards bear their young alive bears little on the dinosaur case except to establish that such a practice *can* occur among reptiles. Crocodiles and turtles, however, lay their eggs. Also, snakes and lizards are only fourth or fifth evolutionary cousins to the dinosaurs, so if internal gestation evolved among dinosaurs, it did so independently.

### What was the contemporary sex life of the average dinosaur like?

Probably better than the contemporary sex life of the average paleontologist. At worst it was clumsy; at best, annual. Judging from the mating habits of crocodiles, the sex life of most dinosaurs was a once-a-year affair or an every-other-year rendezvous. The warm-blooded coelurosaurian dinosaurs, like their avian descendants, may have mated more often.

**Did dinosaurs perform bizarre mating rituals?**

Dinosaurs were by nature into leather, but there's no evidence for *Triceratops* having a horn fetish or for *Brontosaurus* scheduling precoital readings of Proust. Again, birds and crocodiles may provide the best courtship analogies among living animals. Male and female crocodiles in tête à tête are known to bob their heads, undulate their bodies, lash their tails and roar, grunt, squeak and rumble. Crocodiles also use these ritual displays to establish dominance hierarchies—pecking orders to determine who is top dog, or top croc in this case. The birds' version of sexual politics is their well-known peacock prancing, where males strut and display their wares during the competition for females. Among many species of birds, the males and females have elaborate premating courtship dances. It's a safe bet that dinosaurs used sounds and body language to recognize and court mates of the same species, to joust with rival suitors and to settle dominance disputes.

The warm-blooded dinosaurs certainly had the energy and anatomy for hot-blooded courtship. Pachycephalosaurs, the dome-headed dinosaurs, had literal tête à têtes: They probably butted heads, much like bighorn sheep, in the battle for mates and status. Bakker thinks that *Tyrannosaurus* and other carnosaurs also knocked heads in sexual combat.[2]

Hadrosaurs, the duck-billed dinosaurs, had skulls equipped with curvaceous and helmet-shaped bony crests for strutting their sexual stuff and species colors. The crests contained hollow tubes that ran in long loops from the nostrils to the windpipe, allowing the duckbill to honk and trumpet its passion and power.

Of course, what this question really meant to ask is, how did dinosaurs do it? Halstead elaborated on his 1975 reconstruction of dinosaur coitus[3] with five pornographs of Mesozoic matings in the February 1988 issue of *Omni*.[4] The paintings make the passionate contortion, the tangled tonnage of limbs, tails and torsos, look easy: They interrupt a pair of fevered *Tyrannosaurus* and *Brontosaurus* having carnal knowledge in the water and play voyeur with *Diplodocus*, *Brachiosaurus* and *Edmontosaurus* engaging in ecstatic humpery on land. All, according to Halstead, favored rear mounting with "great delicacy and great decorum." But there are few ways to say the following to a child with delicacy and decorum: Dinosaurs, like other reptiles, had no external genitalia; the internal genitalia required of the male dinosaurs in the Halstead-inspired positions defy even the most prurient exaggerations of anatomy.

About as popular as dinosaur sex were questions on the brute strength, feelings, habits and constitutionals of dinosaurs, including two Zen dilemmas:

### Did any dinosaurs shake the earth?

A few years ago, an ambitious physicist with time to kill calculated that if China's 1 billion people jumped in unison off a six-foot height, the collective impact would shake the earth out of its steady orbit. The key here is "in unison." Neither Mao nor all the tea in China could buy such cooperation. Nevertheless, assuming an average weight of 100 pounds per person, the total earthshaking force is 100 billion pounds accelerated by gravity (32 feet/sec/sec) for 6 feet. In dinosaur units, that force translates into 16.7

million *Stegosauruses* (at 3 tons each), or 4.2 million *Diplodocuses* (at 12 tons each), or 50 million *Allosauruses* (1 ton each) or 10 million *Triceratopses* (5 tons each), leaping simultaneously off a six-foot cliff just so they can feel the earth move under their feet.

**Did any dinosaurs crush a pine tree?**

Only after crashing down on one from a six-foot height en route to shaking the earth. Pine trees had evolved by Cretaceous times, the last geologic period of the dinosaurs, and were available for crushing by herds of *Alamosaurus*, *Corythosaurus*, *Tyrannosaurus* and *Triceratops*.

**What kind of dinosaurs lived in Texas?**

Loud, big, bragging ones that shook the earth and crushed pine trees.

**If you were hit by a *Stegosaurus*, how much force would that be?**

Deadly. An adult *Stegosaurus* weighed in at about 4 tons. At an estimated acceleration of 5 miles an hour, the force of impact, assuming an impact time of one tenth of a second, converts to about 17,000 newtons. One newton, named for Isaac Newton, is appropriately the force exerted by the weight of 1 apple resting in the palm of your hand. Being pounded by 17,000 apples is magnum force. At 15 miles an hour, a *Stegosaurus* would whomp you with 41,000 apples. At 35 mph, the stegosaur speed limit, you'd own all the apples in Wisconsin. Physicists in the audience are welcome to submit other calculations.

**Do dinosaurs go to the bathroom?**

Not now they don't, but for 130 million years, during most of the Mesozoic era or the "Age of Reptiles," much of the earth's land surface served as a dinosaur outhouse. The tangible remains of this activity are called coprolites, the official term for fossil dung. An official insult in paleontology is to inform a colleague that his scientific work is best suited for publication in *Acta Coprolitica.* Insults aside, coprolites often preserve the remnants of undigested plants and bones, and reveal the meals that made the coprolite and the diet of the coprolite maker.

**Was *Triceratops* ever scared?**

Well, maybe apprehensive, which is scared with a college education. But *Triceratops* didn't need a college degree to tremble in the ferocious, towering shadow of a twenty-foot, four-ton *Tyrannosaurus* on the hunt. This bloody battle of Cretaceous gladiators has been played out in countless cartoons, movies, illustrations and murals on museum walls. *Tyrannosaurus,* huge jaws agape, six-inch sawteeth bared, circling for a flank attack. *Triceratops,* partially protected by its massive head shield, parrying and lunging, waiting for the opening to charge head down and thrust its long, deadly horns through the soft underbelly of its foe.

I imagine that the line score on these battles is dead even. *Tyrannosaurus* won a few meals. *Triceratops* silenced a few predators. Other encounters were a standoff. Both beasts were the culmination of distinguished evolutionary pedigrees that reached back millions of years in the Cretaceous. Ironically, both were among that handful of dinosaurs that were the last to go extinct and mark the end of an age.

**Did dinosaurs live near Columbus, Ohio?**

Not if they could help it.

**Did dinosaurs live in Pittsburgh?**

Yes, but you'd be hard-pressed to prove it by the fossil record. The terrestrial Mesozoic rocks that could record their presence are rare in Pennsylvania and over much of the eastern United States. But a few bones from New Jersey and Nova Scotia and trackways in Pennsylvania, Connecticut and elsewhere tell us that dinosaurs roamed here.

Only a few of the questions mistakenly put man in the Mesozoic arena with dinosaurs, flying reptiles (pterosaurs), fishlike reptiles (ichthyosaurs) and other extinct beasts from the Age of Reptiles. For example, one asked:

**Can you tell me of any famous people that got eaten by a dinosaur?**

A few were devoured in movies such as *One Million Years B.C.*, which is about 64 million years too late. People (*Homo sapiens*) appeared on earth about half a million years ago, and famous people only since talk shows began, but the last of the dinosaurs became extinct 65 million years ago.

Though not eaten, two famous people were almost pulverized by a dinosaur in 1910. William J. Holland, director of The Carnegie Museum (and chancellor of the University of Pittsburgh), and Arthur H. Coggeshall, the museum's chief preparator, were in St. Petersburg, Russia, supervising the installation of a plaster replica of The Carnegie Museum's famous *Diplodocus* skeleton. By then, Holland and Coggeshall had already mounted five other replicas in natural-history museums in England, France, Germany, Austria and Italy, all without incident. In St. Petersburg, however, in one of the spacious exhibition halls of the Imperial Academy of Sciences, a slapstick mishap almost led to tragic consequences.

The first part of the installation went smoothly. The giant vertebrae of *Diplodocus*'s immense backbone were strung together and strapped to two long steel bars. Then, using block and tackle, Russian laborers began hoisting the steel and plaster mass about fifteen feet off the floor, straining at the guy ropes to keep the colossal backbone elevated. Coggeshall stood on a stepladder under the suspended vertebrae, waiting to help Holland lock the vertical steel support rods in place under the shoulder and pelvic areas of the dinosaur. Holland had picked up the steel supports and was on his way over to the skeleton when, suddenly, the doors to the gigantic hall flew open and a contingent of the tsar's dignitaries marched in. At the sight of royalty, the Russian laborers froze, snapped to attention, and let go of the ropes. Tons of plaster dinosaur crashed to the floor of the academy. The impact rumbled through the building like a small earthquake, rattling the huge portraits of tsars and tsarinas hanging on the walls in an ominous historic portent. Miraculously, the falling *Diplodocus* missed Holland and Coggeshall.

The two men stood in shock amid a thousand pieces of dinosaur vertebrae, a plaster jigsaw puzzle scattered over the floor of the Russian Academy. It was mid-June. They were six thousand miles from Pittsburgh. It would take months to have the museum cast another replica and ship it to St. Petersburg. Instead, Holland and Coggeshall began gluing the pieces together. They worked from dawn into the last of the northern Russian twilight and were done in a week. Then, the following day, as the Russian laborers again began lifting the gigantic plaster backbone into the air, Holland strode over to the entrance of the hall and locked the doors. A month later, in mid-July 1910, the mounted replica of The Carnegie Museum's *Diplodocus* was presented to Tsar Nicholas II.[5]

**Whose idea was it anyway to collect all these bones?**

A question that only a pesky kid would ask. The answer is Andrew Carnegie. While having breakfast on a November Sunday in New York in 1898, Carnegie's attention was riveted by a sensational headline splashed across the front page of the *New York Journal*:

MOST COLOSSAL ANIMAL EVER ON EARTH
JUST FOUND OUT WEST

The newspaper article reported the spectacular discovery of a gigantic dinosaur in Wyoming by a fossil collector named William Reed. The dinosaur's feats were already legend:

> When it walked, the earth trembled under its weight of 120,000 pounds. When it ate, it filled a stomach large enough to hold three elephants. When it was angry its terrible roar could be heard ten miles. When it stood up, its height was equal to eleven stories of a sky-scraper.

Alongside the article was a dramatic scene of a *Brontosaurus*-like beast reared up on its hind legs beside the New York Life Building, its front limbs resting on the seventh floor, its tiny head peering into a window on the eleventh floor. Other shots pictured the colossal skeleton of the dinosaur, its eight-foot thigh bone, its strange skull and its petrified footprint impressions.

No one in New York suspected that the dinosaur "find" was a hoax. Reed had found only a portion of one thigh bone. The drawings of the eleven-story beast were hype—figments of publicity and anatomical extrapolation from other dinosaurs. Carnegie read the newspaper article and was hooked. He penned a note to W. J. Holland in the margin of the paper—"Dear Chancellor, buy this for Pittsburgh"—and sent it to him at the Carnegie Institute (then just a library and music hall) along with a check for $10,000.

Holland assembled the dream team of bonehunters—Jacob L. Wortman, A. H. Coggeshall (and later, John Bell Hatcher and O. A. Peterson)—and sent them west on the *Overland Limited* to meet Reed at his dinosaur site near Medicine Bow, Wyoming. When no other parts of the skeleton were unearthed, a red-faced Reed confessed to finding only the one partial dinosaur thigh bone. Undaunted, Wortman and Coggeshall began to reconnoiter the surrounding badlands. Two months later, in a gully near Sheep Creek, Wyoming, about thirty miles from the site of the "colossal animal," they found the first skeleton of *Diplodocus carnegii*. For me, *Diplodocus* is the measure of the bonehunting profession. It was discovered on Independence Day, July 4, 1899; Wortman and Coggeshall didn't take the day off.

More dinosaurs surfaced in rocky exposures near the *Diplodocus* quarry: *Camarasaurus*, *Stegosaurus*, *Apatosaurus* and a second *Diplodocus*. Over the next three years, hundreds of huge bones of these reptilian behemoths were excavated, encased in a protective plaster and burlap shell, and shipped in wooden crates to Pittsburgh.

Royalty was smitten with *Diplodocus*. During a visit to Skibo Castle, Carnegie's home in Scotland, King Edward VII of England noticed a drawing of *Diplodocus* and asked Carnegie to buy a real one for England. Carnegie relayed the request to Holland, who, aware that the chances of finding another skeleton were slim to none, wasn't about to organize another *Diplodocus* expedition. Instead, Holland answered Carnegie with the polite equivalent of "Get serious," and suggested that a plaster replica of the dinosaur would do King Edward just as well.

A team of skilled statuary plasterers was imported from Italy to make the complicated hollow molds, and Coggeshall devised a method of assembling and mounting dinosaur skeletons on a chassis of forged steel rods, a technique still used today. In 1904, the first cast of *Diplodocus* was erected temporarily in the main building of the Western Pennsylvania Exposition Society in downtown Pittsburgh. At the time, this was the only available building in the city large enough to house the eighty-four-foot skeleton. But not for long. Spurred by the discovery of *Diplodocus*, construction had started on a dinosaur hall at Carnegie Institute. Until the new wing was completed, however, the huge *Diplodocus* molds of backbones, limbs and skull had to be stored in a large brick horse stable behind 419 Craft Avenue near the museum.

A year later, the first replica of *Diplodocus* was formally presented to King Edward VII at the British Museum in London. Not to be outdone, copies were requested by Kaiser Wilhelm of Germany, President Fallières of France, Emperor Franz Josef of Austria, King Vittorio Emanuele III of Italy, Tsar Nicholas II of Russia and King Alfonso XIII of Spain and installed in their respective national museums. *Diplodocus* was the rage of Europe, the "beast which has made paleontology popular."[5]

*Diplodocus* had also launched Andrew Carnegie and his paleontologists on a furious pace of dinosaur collecting that would continue for thirty years. In 1907, Carnegie Institute opened a new wing, a museum of natural history heralded as the "house that Dippy built." Its centerpiece was a dinosaur hall with the skeleton of *Diplodocus carnegii* from Sheep Creek, Wyoming. In 1909, the museum's paleontologists were to discover and excavate a remarkable cache of dinosaurs near Jensen, Utah, including skeletons of *Apatosaurus* (alias *Brontosaurus*), *Allosaurus*, a teenage and adult *Camarasaurus*, *Dryosaurus* and *Camptosaurus*. The site would later become Dinosaur National Monument.[6] In the meantime, *Diplodocus* became the first dinosaur immortalized in a popular tavern song of the day in Europe and America:

Crowned heads of Europe
All make a royal fuss
Over Uncle Andy
And his old *Diplodocus*.

# Chapter 5

*The orangutan, the ape that helped beget Piltdown Man. (From John S. Kingsley,* The Standard Natural History, *Vol. 5:* Mammals, *S. E. Cassino and Co., Boston, 1884)*

# Aromatic Man

As soon as I saw the advertisement for Aramis cologne I smelled a rat. The full-page spread in the Sunday *New York Times Magazine* proclaimed:

> From the Land Where Man was Born.
> Dalla Terra dove e nato l'Uomo.
> A new fragrance for men created by Aramis.
> Firenze. Italia.

The Italian appeals to European chic, the English to those of us with unilingual dollars and scents. The same pitch is made in classy French and Italian restaurants, where the menus outdo foreign-language exams. All this shows is that style sells substance. Fine. But when the Aramis people tell us that Italy is the "land where man was born," the new cologne begins to smell of *eau de bologna.* Italy did bear many great men. Michelangelo. Verdi. Da Vinci. Fermi. Galileo. Columbus. Also Vivaldi, who taught Italians how to sing, and Mussolini, who did so while pounding the lapels of the listener. But no amount of pounding can make man the species call Italy its birthplace, at least if paleontology, not perfume, does the calling.

By Madison Avenue's nonliteral standards, the Aramis people are guilty of nothing more than a clever bit of advertising hype. The same could be said if one called the Magna Carta an Italian credit card. The ad people probably know that from Africa, not Italy, come the extinct ancestral hominids on our evolutionary tree: species such as *Australopithecus afarensis, A. africanus, A. robustus, Homo habilis, H. erectus* and archaic forms of *H. sapiens.* Their petrified skulls, jaws, teeth and skeletons are a bone-hunter's guide to African early-man sites: the Afar, Omo and Middle Awash regions of Ethiopia; East and West Turkana, Kenya; Olduvai and Laetoli, Tanzania; Broken Hill, Zambia; Taung, Swartkrans, Sterkfontein, Kromdraai and Makapansgat, South Africa.

China, Java, India, Algeria, Morocco and Spain also boast ancient bones of *Homo erectus* and *Homo sapiens.* Archaic and modern man in various fossil guises (Neanderthals, Cro-Magnon) hailed from Israel, Iraq, Germany, Hungary, France, Yugoslavia, England and Greece. Also Italy.

More serious than Aramis's hype, though, is the attitude behind the ad. It's apparent the colognizers at Aramis have caught a whiff of "hominid fever," a disease spawned by a driving passion to unearth our own evolutionary roots. The good news about hominid fever is the fruitful chase it leads paleobiologists down the alleys of human evolution. The bad news is the hype, hoax and error the fever can inspire. One symptom is pride in prehistoric pedigree, a Garden of Eden delusion. It makes countries with fossil hominids feel somehow anointed. This odorous sort of anthropological nationalism is the honor Aramis claims for Italy and Aromatic Man.

Along with a fever, fossil hominid discoveries evoke our own *terra incognita,* as if they held the harbingers of our primeval biological fate. Understandably then, even deservedly, they can create a peculiar glamour. The petrified bones and teeth of our ancestors made household names out of Leakey, Johanson, Peking Man and Neanderthal, and best sellers out of *Origins* (by Richard Leakey and Roger Lewin) and *Lucy: The Beginnings of Humankind* (by Donald Johanson and Maitland Edey), not to mention Jean Auel's fictional portrayal of Pleistocene passions in *The Clan of the Cave Bear.* I doubt whether the nonhominid equivalent—say, *Minnie: The Beginnings of Rodentkind*—would have topped the *New York Times* best-seller list, although rodents predate hominids by some 50 million years.

Fortunately, most students of human origins have resisted deleterious doses of hominid fever. But a few become infected, one as early as 1726. At that time, scientific opinion in Europe had come to recognize fossils as evidence of past life. Originally, the word "fossils" meant "objects dug up," all of which were considered inorganic and lumped with minerals, along with crystals, ores and useful rocks. Fossils were not identified as petrified remains of ancient organisms until the distinction between living and inert objects became clearer, and the notion of eras of past life accepted. But how past was "past life" to an early eighteenth-century naturalist? The prevailing view accommodated scripture: Fossils represented pre-Diluvian life; they had been laid down and buried by the worldwide Deluge.

It was in this scientific atmosphere that a Swiss naturalist and physician, Johann Scheuchzer, was sent a fossil skeleton uncovered by workmen in a limestone quarry near Oeningen, Switzerland. Scheuchzer had long sought human fossils as testimony for biblical truth. In 1726, after sketching and studying the skeleton, Scheuchzer proclaimed it as unmistakable evidence of the Flood—the body of a poor sinner who had perished in the Deluge. He named it *Homo diluvii testis*, "flood man."

Sadly, Scheuchzer's zeal for human relics of the Deluge

had blinded his knowledge of anatomy. His *Homo diluvii testis* was not a fossil sinner but a fossil salamander. The error was discovered eighty-five years later, in 1811, by Georges Cuvier, the eminent French paleontologist and naturalist. The Oeningen skeleton, now known to be about 10 million years old, belongs to a species of salamander (called *Andrias scheuchzeri*) that survives today in China.

Diluvian notions weren't to blame for one of the more recent attacks of hominid fever. In 1979, Noel Boaz, an anthropologist then at New York University, announced the discovery of a partial clavicle (collarbone) of a small, primitive hominoid, or ape-man, at Sahabi, a five-million-year-old site in Libya. The excavation also yielded remains of various marine animals (sharks, turtles, sea cows) and a host of terrestrial birds and mammals.

In 1982, in an article in *Natural History*, Boaz reconstructed the ancient Sahabi scene: grasslands, swamps, woodlands, lush forests. The anatomy of the clavicle, he concluded, indicated that the ape-man got around by "swinging arm over arm as apes do today." Ten months later, Tim White, an anthropologist at Berkeley, and his colleagues tossed the swinging Sahabi "ape-man" into the drink. According to White, Boaz had misidentified the collarbone: It was really part of a rib of a fossil whale.

Even paleontological savants can succumb to hominid fever, especially if the "find" occurs in North America, a continent devoid of fossil anthropoids—extinct monkeys, apes and man. Such a discovery would be momentous. And it was. In February 1922, a consulting geologist and paleontologist named Harold Cook found a large, worn, oblong tooth in 10-million-year-old rocks near Snake Creek, Nebraska. Cook thought the tooth looked humanlike. He sent it to the eminent paleontologist Henry Fairfield Osborn, who was president of the American Museum of Natural History in New York, and a past curator in its paleontology department. Two months later, after comparing the tooth to those of other fossil animals, Osborn agreed with Cook. He heralded the tooth to the scientific world as an upper molar of "the first anthropoid primate found in America," and named it *Hesperopithecus* ("western ape") *haroldcookii*.

Other paleontologists at the American Museum of Natural History concurred: The tooth closely resembled an upper molar of the extinct Java ape-man, *Pithecanthropus erectus* (now called *Homo erectus*). But critics and skeptics in the scientific community submitted a litany of other possible identifications. The fossil tooth, they said, could also be the first upper milk tooth of a fossil horse, or the middle ear bone (incus) of a gigantic fossil mammal, or the molar of a fossil bear, a giant raccoon or a giant panda, or the tooth of a gigantic relative of a South American monkey. All of these suggestions were refuted. But it just goes to show that a large, worn, oblong tooth is not the Rosetta Stone of paleontology.

*Hesperopithecus* survived as an anthropoid until 1927, when expeditions to the same Snake Creek beds turned up jaws with worn and unworn teeth of a fossil peccary (*Prosthennops*), a piglike animal. The worn molars matched the tooth of *Hesperopithecus*, ending the notion of a "western ape." The fervent desire to discover a North American fossil hominid had, temporarily, made a primate out of a peccary. As the British anatomist Grafton Elliot Smith remarked at the time, paleontologists had been badly bitten by the Nebraska tooth.

Smith was a fine one to talk. At that moment he was being bamboozled by the greatest hoax in human paleontology—Piltdown Man. Joining him as dupes were Arthur

Keith, also an anatomist, and Arthur Smith Woodward, a paleontologist at the British Museum of Natural History. Smith, Keith and Woodward were the brain trust of human evolution in Britain in the early 1900s and unwittingly promoted the hoax.

In 1912, Britain, like North America, had no homegrown fossil hominids to count among the treasures of the Empire. By December of that year it did, courtesy of Charles Dawson, a Sussex solicitor and amateur geologist who combed the local gravel pits for fossils. A few years earlier, workmen at one such pit near Piltdown Common had come across a few pieces of humanlike skull bones. They gave them to Dawson, who took them to Woodward at the British Museum. Both returned to the Piltdown pit in the summer of 1912 for serious excavation, and were joined by two French priests then studying at the Jesuit college near Hastings. One of the priests was Father Pierre Teilhard de Chardin, who later became somewhat of an authority on human evolution and, in some quarters, one of the prime suspects in the Piltdown hoax.

The pit produced more brownish skull remains and part of a lower jaw with two worn teeth. Also recovered were remains of elephant, mastodon, red deer, beaver and horse; they indicated an early Pleistocene age, about 2 million years, for Piltdown. But most of the Piltdown hominid was missing: half the skull, the chin, the jaw joint, all but two teeth, and the skeleton.

Nevertheless, Woodward pieced together and reconstructed the entire skull, lower jaws and teeth, Smith analyzed a cast of the brain cavity, and on December 18, 1912, before the Geological Society in London, Piltdown Man was officially born. They called it *Eoanthropus* ("dawn man") *dawsoni*. It was the perfect missing link between humans and apes, with a humanlike skull, apelike lower jaws and teeth, and according to Smith, a smallish, primitive brain.

At first Keith dissented, claiming that Woodward's reconstructions of the missing skull bones and teeth were too apish. His own version gave *Eoanthropus* a more human countenance. But in the next three years more skull bones and apelike teeth surfaced at the original Piltdown pit and at a new site nearby. Keith conceded. *Eoanthropus*, allegedly from the early Pleistocene, became Britain's candidate for the oldest known ancestor of modern man. Later, it helped propel Smith, Keith and Woodward to knighthood.

But critics in America and Europe were skeptical. What bothered them was the anatomical paradox of an apelike jaw with a humanlike skull in one individual. Apparently, none of them had ever met Mussolini. Marcellin Boule, a French anthropologist, and Gerrit Miller, an American mammalogist, put it bluntly: The skull was human, the jaw a chimp's, and Piltdown Man was a composite. They were overruled by the scientific prestige of the British triumvirate, and by the unshakable evidence of association—the jaw and some of the skull bones had been found in the same pit alongside extinct mammals.

The geologic age, authenticity and enigma of *Eoanthropus* remained unchallenged for almost three decades until, in 1948, Kenneth Oakley, a geologist at the British Museum, applied his newly developed technique of fluorine dating to all of the Piltdown fossils. The remains of the extinct animals came up early Pleistocene, but the *Eoanthropus* bones registered post-Pleistocene—less than 10,000 years old. Something was rotten at Piltdown, and Joseph Weiner, an anatomist at Oxford, smelled a fake.

The ape-human anomaly of *Eoanthropus* had always been fishy. Now there were two more discrepancies. First, bones from the same rocks in a pit should date to the same geologic age. Second, if Piltdown Man was the "missing link," how could it be no older geologically than its supposed descendant, modern man?

Investigators had only been allowed to examine plaster copies of the original Piltdown fossils, but in 1953, Weiner received permission to examine the real bones and the jig was up. The hoax was ingenious. The apelike teeth had been filed down to remove critical anatomical clues; the most diagnostic portions of the lower jaw—the chin and joint area—had been deliberately broken off and removed; all of the bones had been stained to match the color of the Piltdown rocks. The skull was human, the jaw an orangutan's. Both were modern, and both had been planted at Piltdown.

But whodunit? And why? All the players are suspect, but, recently, two have been indicted by motive, opportunity and circumstantial evidence:[1] the amateur Dawson and the priest Teilhard de Chardin, either acting alone or as co-conspirators. Both worked the Piltdown site initially and found almost all of the *Eoanthropus* "remains." As for motive, Dawson may have wanted renown and respect to counter the "Sunday geologist" reputation he had at the British Museum. Teilhard may be guilty of nothing more than playing an intellectual, good-natured, evolutionary prank, which, to protect his scientific reputation, he could not retract once the "discovery" got out of hand.

Even Arthur Conan Doyle, who lived a few miles from the Piltdown site, has been singled out as the perpetrator.[2] With his gift for mystery plots and his proficient knowledge of chemistry, anatomy, paleontology and anthropology, Doyle certainly had the talent to carry it off. But the evidence against Doyle is as thin and circumstantial as that against Dawson and Teilhard. Doyle apparently visited the Piltdown site after the discoveries were announced, corresponded with Dawson about the "skull" and once excavated dinosaur footprints with Woodward. His novel *The Lost World* contains some vague parallels with the Piltdown affair although it was conceived, written and published before the 1912 announcement of Piltdown Man. Doyle also had access to human jawbones and, perhaps, the animal fossils and flint tools planted at Piltdown—he and his wife honeymooned in the same Mediterranean areas where many of those artifacts and fossils may have been originally collected.

If Doyle had opportunity and method, what was his motive? Perhaps to humiliate the British scientific establishment, especially Edwin Ray Lankester, who, until 1907, was director of the British Museum of Natural History. Lankester, a staunch advocate of science over pseudoscience, often published vitriolic attacks on Britain's Spiritualist movement of which Doyle, for all of his scientific acumen, was an avid acolyte. With a monumental scientific swindle, such as the one at Piltdown, Doyle could embarrass Lankester and burst the pompous balloons of the British Museum. The flaw in this thesis is the unconsummated finale to the hoax: Conan Doyle never came forward (as his pontifical Sherlock Holmes invariably did) to proclaim his deception publicly, make Lankester a laughingstock and bask in the limelight of sweet revenge.

More important than whodunit is the reason for the hoax's success: hominid fever. When living man studies fossil man, ego and fancy can perfume the facts. In the case of the Piltdown hoax, hominid fever blessed the marriage of shoddy evidence to willful expectation. The same union sired "flood man," "western ape," the Sahabi "hominoid" and Aramis's Aromatic Man. It's well then that we heed the paleocratic oath: Old bones, like perfume, should be sniffed before they're swallowed.

# Chapter 6

Bison charging, painted by a Palæolithic artist on the wall of a cave at Altamira.

Sometimes this is described as a bison lying down. But who ever has seen an animal repose with its tail cocked up in the air?

*Art in the Ice Age. Were the Cro-Magnons painting the Pleistocene fauna or just making a Freudian statement? (From Keith Henderson,* Prehistoric Man, *Chatto and Windus, London, 1927)*

# Art on the Rocks

To understand art one has to know French. Most useful is *oeuvre*, a fancy word for an artist's collected works. *Oeuvre*, however, shouldn't be confused with *hors d'oeuvre*, the fancy tidbits served at openings of art exhibitions, or with the Louvre, the great art museum of France, where the *oeuvre* and *hors d'oeuvre* are the *crème de la crème*.

Another handy phrase is *je ne sais quoi*—literal translation: "I don't know what." Critics use it diplomatically—"It has . . . what shall we say . . . a certain *je ne sais quoi*"—to praise the creative intangible when confronted with creative rubbish. There's also *genre*, to denote a discrete artistic style, and *avant garde*, to describe a new, bold experimental *genre*, usually after the experiment has failed.

A successful opening is a *tour de force*, but if one finds an art show tedious, it's more *chic* to complain of *ennui* than boredom. These *bons mots* and others make it clear that in art, French is worth a thousand words.

It's fitting then that much of man's earliest art comes from France. Thirty thousand years before the Louvre hung any *oeuvre*, Ice Age artists—Cro-Magnons and other early *Homo sapiens*—began painting and engraving the walls of limestone caves and rock shelters in southwest France, northern Spain, Italy, north Africa, Tanzania, and South Africa. It was an artistic explosion that reached its height about 15,000 years ago and faded with the retreat of the last Ice Age 10,000 years ago. Today, each of the more than two hundred known painted caves and rock shelters is an Ice Age art and natural-history museum: Lascaux, La Mouthe, Font-de-Gaume, Les Combarelles, Le Tuc d'Audoubert, Les Trois Frères, Pech-Merle, La Marche, Altamira—the first painted cave to be discovered—and others. Among these, France's Lascaux and Spain's Altamira are the Louvres and Smithsonians of ancient man.

Galloping over these cave walls and ceilings are herds of red and black horses, bison, deer, oxen, mammoths and ibex. Some are swollen in pregnancy. On some of the rock faces prowl lone yellow lions, hyenas, foxes and wolves amid the brown reindeer, antelope, boars, goats and rhinoceroses. A few fish and birds round out the painted bestiary. Some of the walls are a lesson in order and composition. Others are a jumble of images superimposed on the same rock, perhaps due to the work of successive artists or a paleo-impressionist's attempt at a three dimensional scene.

The Cro-Magnon palette was made up of red, black, yellow and brown mixed from natural clays, mineral oxides, charcoal and animal fat; the artists applied the paint with their fingers, clumps of moss, twigs and primitive brushes, or sprayed it through hollow reeds. Some of the animal images are simple outlines incised on the cave wall or carved in bas-relief. A few are sculptures, molded from lumps of clay and ground bone and fired rock-hard in ancient kilns. Cro-Magnons also had their version of traveling exhibitions—archaeologists call it "portable art": pieces of antler, ivory and bone engraved with animal figures and worn smooth through years of handling.

The techniques of the Cro-Magnon artist may be clear, but their motives are not. Why did they paint, etch and sculpt a menagerie of Ice Age beasts throughout much of the Old World? Art may reflect life, modern or ancient, but like art today, interpretations of Ice Age *oeuvre* are as given to claptrap as consensus.

The search for the Ice Age artistic catalyst should begin with the life of the artist: the environment and ecology of the Cro-Magnon people. As nomadic hunter-gatherers, they lived in bands of twenty-five to thirty individuals and inhabited caves, rock shelters, freestanding huts or pits in the ground, depending on the season and climate. During the late Pleistocene, from 30,000 to 10,000 years ago, much of Scandinavia and England was covered by glacial ice. Most of Europe was a treeless tundra, save for the forested mountain slopes and valleys. Summer temperatures in southern France and Spain averaged 59 degrees, temperatures that today would cause mass extinction among the trendy on the Riviera. Their winters made Buffalo seem balmy.

Whatever other motive is due the cave art, one surely was to imitate life, to render a checklist of the wild creatures, perhaps the game, of that era and locale. Corroboration comes from the fossil record, which says two things about the wildlife outside the caves. First, game was plentiful. Second, the species that roam in paint across the cave walls also roamed Ice Age Europe.

If the art of the Cro-Magnons was dazzling, so was their technology. Instead of being made only of flint, the mainstay of the earlier Neanderthal toolmakers, more than half the tools in the Cro-Magnon "tool kit" were fashioned from new, stronger and more workable raw materials—bone, antler, ivory. Archaeological excavations have uncovered a veritable hardware store of implements: knives for cutting meat and whittling wood, scrapers for cleaning bones and hides, perforators, picks, burins, chisels, harpoons, shovels, stone saws, pounding slabs and more. Flint axes and knives were fitted with bone and antler handles for a firmer grip and better leverage. Other inventions were a bone-handled spear-thrower and barbed spearpoints.

Given this sophisticated gear for slaughter and dismemberment, it's no wonder that Cro-Magnons were hunters *par excellence* who knew the behavior and habits of their prey. Ice Age sites in Czechoslovakia and France preserve the remains of hundreds of mammoths lured into pitfalls, and thousands of horses either deliberately stampeded off a cliff or cornered in a box canyon.

From this context come the three prevalent anthropological interpretations of man's earliest art:

1. Symbolic or magical prehunting ritual
2. Symbolic sexual duality
3. Symbolic female-centered society

Predictably, the emergence of the three schools followed twentieth-century societal fads. The first coincided with traditional notions of "Man the Hunter-Provider," the second with the ascent of Freudian psychology, and the third with the recent rise of feminism. The only consensus is that there was no art for art's sake among the Cro-Magnon. Rather, their *oeuvre* is pregnant with symbolic meaning. If so, some of the inferred meanings are premature, if not miscarriages.

Proponents of the prehunt ritual maintain that the images were executed in a symbolic ceremony to ensure the success of the hunting party. Evidence brought to bear are the numerous cave paintings of speared bison, horses and mammoths, and the geometric designs on the rocks that can be taken to be images of snares and traps. Also, similar rituals occur among some living groups of hunter-gatherers.

But the prehunt-ritual interpretation may be somewhat simplistic. Archaeological remains show that reindeer were the most commonly eaten animals, yet their images are extremely rare. Horses, bison and oxen, the animals most frequently depicted, apparently appeared *à la carte* much less often. Some modern hunter-gatherers, such as the San of Namibia and Australian aborigines, draw certain animals that are "good to think" rather than good to eat.

A prehunt ritual also does not explain the peculiar distribution of the art in the caves, almost as if the paintings had been hung by a committee of art curators. For example, much of the cave art is in the large galleries near the cave entrance and natural light; but many of the spectacular animal portraits were painted in the darkest and least accessible recesses of the caves, where there was room for only the artist and a few moss-wick lamps.

More cogent is Louis Leakey's variation on the theme of ritual hunting.[1] Leakey maintains that a good deal of the art in the open galleries is commemorative, a record of observations and deeds. The animals were painted after being sighted, stalked, admired, captured, missed or killed. Some of the portraits, and especially the sculptures, were simply home decorations or unabashedly art for art's sake.

Of most "magical" connotation, according to Leakey, are the animal images in the hidden niches of the caves. They represent the symbolic "tying up" of wounded game that a hunter maimed and tracked but didn't capture. This ritual, practiced until recently by a number of African tribes, symbolically "safeguards" the wounded animal from escaping or falling prey to other hunters. "Tying up" the animal in the least accessible part of the cave was an additional "safeguard" to keep it out of sight of any other predator. Similarly, painting pregnant animals helped, symbolically, ensure the continued fertility and abundance of the game.

The second—Freudian—interpretation of cave art abandoned hunting for sexual symbolism, and was arrived at independently by two French prehistorians. According to them, the animals depicted and their distribution within the caves mirrored male/female duality in Cro-Magnon society. Trouble is, the two prehistorians came up with a sexual identity crisis: One of them said bison were symbols of Cro-Magnon females and horses of males; the other prehistorian concluded the opposite—the horses were Cro-Magnon females and the bison males.

I can't buy it. Reason, like art, consists of drawing the line somewhere. Why in blazes make Freud retroactive to the Pleistocene? What would possess the Cro-Magnons to ascribe "femaleness" to a male bison and "maleness" to a pregnant mare? Certainly not Oedipus. For one, he came much later than the Cro-Magnons; for another, he was in love with his mother, not a bison or a horse. If they wanted to depict "sexual duality," why didn't the Cro-Magnons just portray themselves?

Well, the third interpretation of Ice Age art not only buys the bison-as-female and horse-as-male symbolism, it extends it to other animals. Deer and boar apparently also represent Cro-Magnon "males," and certain groupings of animals in the caves supposedly reflect the social organization and behavior within and among Cro-Magnon bands. An example is one of the main scenes at Altamira, which shows a group of bison, perhaps in the act of giving birth, surrounded by a horse, a wild boar and a female deer. It is taken to mean a matrifocal society in which the females occupy a central role and the males, with their far-flung hunting activity, a more peripheral position.

This interpretation too has a certain *je ne sais quoi.* One expects this kind of symbolism more from the Cannes Film Festival than from a Cro-Magnon cave. Experts less given to symbolic Freudian psychobabble don't see the bison as giving birth, but in the midst of a full charge or in the final throes of death.

It is not the bison but the horse *d'oeuvres* that cracks an intriguing window into the possible practices of the Cro-Magnons: animal domestication. The horses are richly portrayed, accurate and most common. Obviously, they were the favorite subject of Cro-Magnon artists for aesthetic or deeper reasons, one of which may have been horse domestication. Three lines of evidence imply that the Cro-Magnons may have been practicing horse husbandry as early as 30,000 and certainly by 15,000 years ago.

Cowboys and veterinarians will tell you that tethered, corralled or stabled horses develop a condition called "crib-bite," or "cribbing," from chewing or gnashing their lower front teeth against hard objects, usually fence posts and stable doors. Apparently they do it out of nervousness and horse *ennui.* In any event, the tips of their incisors wear into a chisellike shape, whereas those of wild horses remain flat. Two Ice Age sites in France, both dated to about 30,000 years, have yielded some horse teeth showing crib-bite.

Stronger implications of some form of horse domestication come from two 15,000-year-old caves in France, La Marche and Le Placard. Two horse heads, one carved and the other engraved, bear markings that look tantalizingly like a harness. Also, a few of the cave paintings of horses show harnesslike lines across the muzzle, and some of the geometric images on the cave walls may depict animal pens. As one British archaeologist commented, the "Cro-Magnons learned to do more with horses then simply throw a spear at them."[2] Tethered horses may have been ridden, kept as pets, used as decoys in hunting or as beasts of burden. Their constant presence near a campsite may explain their common presence on cave walls.

Images of humans are as sketchy and scarce as those of horses are pervasive and vivid. Most of the human images are simple stick figures, with little or no facial detail. Only one cave, La Marche, features more humans than animals. Here, thankfully, there is no symbolism: Females are females, males are males, and the horses and bison roam free of their Freudian ids. One of the Cro-Magnons is shown with a headband, and several with headgear. Ten of the men have beards, and three, moustaches. In Tanzania too, Stone Age paintings depict elongate, spidery individuals hunting and dancing across the rock shelter. Perhaps Ice Age people had a taboo against the realistic "capture" of the human form and face—a stricture common to many cultures.

If so, the taboo did not extend to carved statuettes of the female Cro-Magnons, of which hundreds have been recovered from archaeological excavations in western Europe. They are popularly known as Venuses: The exaggerated bosom and buttocks are purportedly an allusion to fertility. Actually, only a few of the Venus statuettes qualify as "mother-goddesses." Most of the female figurines are merely broad in the beam, an indication, perhaps, that beauty to the Cro-Magnons was a zaftig torso. Fifteen thousand years later Rubens paid similar homage to chubby feminine charm.

For that matter, much of cave art would find kinship in the Louvre. The animal paintings evoke primitivism, some of the bison, pointillism, and the gaunt human images, Giacometti sculptures. In fact, if faced with one of Giacometti's bronze stick figures, I imagine an Ice Age art critic would have shrugged, and remarked: "Derivative, but not uninteresting."

# CHAPTER 7

*An "Orang-Outang" creation story: in the Garden with tree, staff and pear. (From E. Donovan,* The Naturalist's Repository, *Donovan, Simpson & Marshall, London, 1823)*

# All About Eve

In the beginning, says Scandinavian legend, was an abyss. To the north lay a chaos of ice, darkness and mist; to the south, the land of fire. One day, warm winds from the land of fire brought the first spring and slowly thawed the ice. From the melt waters arose Ymir, a frost-giant of heroic human form, and Audhumla, the cow, whose milk gave Ymir strength. Two more frost-giants sprang from the sweat of Ymir's body and another arose when Audhumla licked the ice. But when the frost-giants multiplied they also bore three gods, and the giants and the gods fought for empire of the universe.

The gods slew the giant Ymir and fashioned the world from his body. From Ymir's flesh they made earth; from his blood, the oceans and rivers; from his bones, the mountains; from his teeth, the rocks and stones; from his hair, the plants and trees; from his brain, the clouds; and from his skull, the vault of heaven. Ymir's eyebrows became the ramparts to the abode of the gods, which lay in the Midgard ("middle land") between the land of icy silence and the land of fiery sun.

The world was now ready for mankind. It was enveloped by a majestic ash tree of perpetual life, Yggdrasil, which reached all lands, sheltered all animals and sent roots into the Fountain of Youth and the Fountain of Knowledge. Finally, from the trunks of an ash and an elm tree that washed up on the seashore, the gods fashioned Askr and Embla, the first man and woman.[1]

Creation tales are alchemy. They unfold magical maps and weave enchanted trails back through wonderlands to the first epic humans. The tales crisscross folklores, the product of multicultural roadwork. Trees of life and knowledge spring up on the garden path as a fable snakes back through Eden to Eve and Adam and clay. Aboriginal Australians, New Zealanders and Melanesians speak of Pund-jel, Qat and Tiki, supernatural beings who fashioned men and women out of red clay and willow twigs. African and American tribal tales bear the first humans from wondrous trees, plants and rocks. One Greek legend recounts how men and women arose from figures baked in clay by Prometheus.

Creation tales are storybook. They have charmed children ever since Adam wore short pants. They flatter adults, tracing our ancestry back to ancestors better than we are. Sometimes they tempt us into simplistic realism, to cross the magical borders of the storybook. In one sad example, during February of 1988, former astronaut and moon visitor James Irwin went looking for the remains of Egyptian chariots at the bottom of the Red Sea. Irwin hoped to stamp the imprimatur of science on biblical passage and Passover plot, but the search came up empty. So too have his six expeditions to Mount Ararat in Turkey to recover the pieces of Noah's ark.

Perhaps he chose the wrong scripture and the wrong mountain to bless with scientific credence. According to Greek legend, Zeus sent a deluge unto Earth to rid it of a violent and vicious race of humans called "the men of bronze." Zeus's flood spared two people, Deucalion, son of Prometheus, and his wife, Pyrrha, who built an ark and drifted for nine days and nights until the floodwaters receded. The ark landed in Greece, not Turkey, and on Mount Parnassus rather than Ararat. Upon disembarking Deucalion and Pyrrha were instructed by Zeus to repopulate the world by casting over their shoulders rocks and stones, "the bones of Mother Earth." Those tossed by Pyrrha turned into women, Deucalion's into men.

Evolutionary tales are also magical maps. They unfurl time, coursing back into extinct eras, over exotic continents and past fantastic bestiaries. But, unlike creation stories, they are charted according to tangible signs: the structure of the universe; the motions of earth and sun; gritty rocks, petrified bones, faunas, floras, molecules, men, birds and beasts. Evolutionary maps also grow and age and acquire new lines, their topographies being ever redrawn to fit the shifting, spreading terrain of evidence, inquiry and discovery.

The traditional evolutionary paths to the first humans are flagged with the fossils of early man in Africa, Asia, Europe and the Middle East. They stretch 2 million years into geologic time to primitive, extinct species of *Homo: Homo habilis* and *Homo erectus*. They wind past early, scattered, diverse populations of *Homo sapiens* dubbed ante-Neanderthals, Neanderthals, archaic *sapiens*, proto–Cro-Magnons, Cro-Magnons and "anatomically modern humans" who lived between 10,000 and 500,000 years ago and are distinguished from one another by anatomical minutiae of the skull and dentition.

But there is a new and innovative evolutionary route to early man, marked by microscopic clues: the chemical dots and dashes of the genetic code strung along the DNA chain in human cells. This molecular pathway has been blazed by geneticists and biochemists, most recently by three researchers at the University of California, Berkeley. Their conclusions hark back to legend, to a primordial mother Eve and a proverbial Eden: Modern humans stem from one woman who lived about 200,000 years ago in Africa.[2]

The research weapons of the Berkeley scientists are that famous duo of heredity and statistics. Heredity is something most parents believe in, at least until their children start acting like fools. And statistics, cynics are fond of saying, is the science of transforming unreliable figures into reliable facts. The so-called "mother Eve" deduction follows from a statistical analysis of part of the hereditary DNA code of 147 people from five geographic populations—Africa, Asia, Australia, New Guinea and Europe/Middle East. The DNA molecules in question aren't the familiar genes and chromosomes in the nucleus of cells. Rather, the Berkeley study focused on different cellular structures called mitochondria, the tiny amplifiers that power the metabolic activities of cells. Mitochondria also carry their own DNA, called mitochondrial DNA, completely separate from the DNA in the nucleus.

Mitochondrial DNA and nuclear DNA are not only in different parts of the cell, they are inherited in fundamentally different ways. The DNA in the nucleus—the genes that determine physical traits—comes half from the mother, half from the father. In contrast, the DNA in mitochondria is strictly maternal, being transmitted only in egg cells, not in sperm cells. Children inherit all their mitochondrial DNA from their mother, their maternal grandmother, the maternal great-grandmother and so on. Mitochondrial DNA is a purely matrilineal genetic legacy, marking a trail of descent that only women walked.

The Berkeley scientists mapped this ancient DNA trail by using special enzymes to chop each of the 147 strips of mitochondrial DNA into discrete segments. Comparison of the segments revealed 133 basic types of mitochondrial DNA. The differences from one type to another lie in the sequence and proportion of constituents, of the individual dots and dashes making up the DNA chain. These differences arise from mutation—random copying errors in the genetic code that can occur every time the mitochondrial DNA reproduces itself in the formation of a new cell. Mutation also afflicts DNA in the nucleus, but the mutation rate of mitochondrial DNA is ten times higher than that of nuclear DNA—about 2 percent to 4 percent of all the mitochondrial DNA will mutate every million years.

As a result, mitochondrial DNA is theoretically both a genealogical beacon and a molecular evolutionary clock. Genealogy is determined from the oldest and ancestral type of mitochondrial DNA, which should have suffered the greatest amount of mutation and should exhibit the greatest diversity in its genetic code. The molecular clock can tick off the amount of time it took for those mutations to occur.

When the 133 different types of mitochondrial DNA were mapped, grouped and ranked according to the five geographic populations, the African sample showed the most diversity and the greatest disparity from the other samples. The African group, concluded the Berkeley workers, is the oldest lineage and at the base of the human family tree, followed by the Caucasian-Mongoloid groups. But how old? If the mutational clock ticks at a constant rate of 2 percent to 4 percent per million years, the African type of mitochondrial DNA took 140,000 to 290,000 years, or an average of 200,000 years, to evolve its unique genetic code. Clearly, then, say the Berkeley biochemists, the ancestry of modern humans can be traced back along a maternal lineage of mitochondrial DNA to a genetic mother Eve who lived in Africa about 200,000 years ago.

Well, the Berkeley team doesn't know mother Eve from Adam's old ox, says a competing group of geneticists from Emory University. Their research indicates Eve didn't live in an African Eden but an Asian one, a conclusion arrived at after similar finagling of mitochondrial DNA sequences, with some novel twists. Their samples of mitochondrial DNA came from 700 individuals on four continents and was chopped by enzymes into larger pieces. Unlike the Berkeley study, the Emory University team found that human races exhibit distinctive types of mitochondrial DNA. The oldest pattern and the one most similar to living apes (and our putative ancestors) apparently occurs in Asia. Statistical treatment of this mitochondrial DNA data generated an evolutionary tree rooted in a rival Asian Eve who lived in south China between 150,000 and 200,000 years ago.[3]

Cynics have a second definition for statistics: fiction in its most uninteresting form. To go the cynics one better, it seems from these mitochondrial DNA studies that statistics, like witnesses, will testify for either side. To complicate matters, the Emory University group claims its samples are more reliable because the Berkeley study used American Blacks as a source for African mitochondrial DNA. The Berkeley team counters that its DNA-splitting techniques are more precise and yield DNA sequence maps of finer resolution.

But the problems of mother Eve extend beyond statistics and methodological quibbling between research teams. A critical issue is the assumptions inherent in the research. First, does mitochondrial DNA mimic a molecular clock, mutating at a constant rate? If so, does it mutate at a constant rate in all human populations? Second, could more than one woman have had the same or very similar ancestral mitochondrial DNA sequences? Perhaps there were two, ten or ten thousand Eves? Third, the statistical manipulation of the mitochondrial DNA data yields a forest of evolutionary trees, ranging from those that require the fewest mutations to those laden with repeated, independent mutations in different evolutionary branches. Which tree is correct?

**Assumption 1: Constant molecular clock.** Without the assumption of a constant rate of mutation in mitochondrial DNA, a 200,000-year-old mother Eve would quickly fall from grace. Some biochemists defend the notion of regular clocklike mutation. Others denounce it: "There is at present no convincing evidence that mitochondrial DNA evolves like a molecular clock, at rates that would persist from one group of organisms to another over long evolutionary spans."[4] These naysayers have made the case that mutation rates can vary with time, with the organism and with the particular segment of DNA.

For example, much of the DNA code is silent or redundant—the genes are seemingly mute—whereas other sequences along the DNA chain are verbose, dictating the synthesis of proteins and enzymes. The silent portions of the code appear to mutate at a relatively constant rate, the active genes at wildly different rates.[5] Furthermore, the mitochondrial DNA mutation rate also seems to depend on the animal: It is ten times higher in some mammals than in others.[4] Perhaps diverse rates apply to human populations as well. The upshot is that the molecular clock may be too mercurial to tell evolutionary time or date a mother Eve at 200,000 years.

**Assumption 2: Two, ten or ten thousand Eves.** Equally uncertain is the assumption that only one woman carried the ancestral sequence of mitochondrial DNA in her cells. Data from well-studied organisms, such as fruit flies, indicate that the same mitochondrial DNA sequence can occur in as many as 44 percent of females at any one time. In other words, there may have been hundreds or thousands of Eves with the same or similar "correct" ancestral mitochondrial DNA.

**Assumption 3: The simplest of evolutionary trees.** Compounding the mother Eve matter is one of the basic doctrines of the scientific method. Scientists shave with Occam's Razor, a principle of logic formulated by William of Occam in the fourteenth century. The principle holds that explanations of phenomena should not be more complex than necessary. In other words, given a choice, scientists should opt for the simplest solution, a tenet also known as the principle of parsimony. In the case of mother Eve and mitochondrial DNA, the simplest solution is an evolutionary tree requiring the fewest mutational steps.

But is nature always most parsimonious? A chorus of evolutionary studies says no. Throughout earth history, unrelated organisms have repeatedly and independently evolved similar physical attributes and anatomy, including DNA sequences. The bane of Occam's Razor is scientists with beards. The bane of parsimony is evolutionary convergence and chance.[6]

Given the degree of repetition and redundancy in the DNA code, I would venture that mutations in mitochondrial DNA are one of those phenomena that William of Occam would have excused. In any case, only a few mutational steps separate the African, Asian, European, Australian and Middle Eastern patterns of mitochondrial DNA. The same mutations could well have occurred independently in different lineages in different geographic areas. If they did, the DNA sequences observed today would have resulted from both genealogy and independent random evolution and could not be backtracked through maternal dowries to a single female.

Assumptions aside, another problem with mother Eve involves competing evidence: How does the trail of mitochondrial DNA square with the traditional fossil record of human evolution? Which of these paths most faithfully traces the history of our species? Paleontologists don't come to this debate dressed like Adam. Not only are their pockets stuffed with the fossil skulls, teeth and skeletons of early man, their clothes are colored by different notions of evolutionary tempo and different interpretations of the anatomy of the ancient bones. There is consensus that *sapiens* and *sapiens*-like fossils extend the human trail back perhaps 500,000 years to a connection with *Homo erectus*, a species (Peking Man, Java Man and others) that appears to have emerged from Africa and spread to Asia about a million years ago and to Europe possibly half a million years later.

But paleoanthropologists do not agree on the *erectus-sapiens* transition. Some interpret the fossil evidence to say that the transition occurred independently in different parts of the Old World, with sporadic mixing of the different lineages. This view cites a diversity of fossil peoples—Neanderthals, Cro-Magnons and others—inhabiting specific geographic regions in Africa, Europe, Asia and the Middle East and differing from one another much as modern peoples of the world do. These semi-isolated populations may have been the progenitors of the modern "races" of humans.

Others contend the *erectus-sapiens* transition occurred at least 100,000 years ago in one geographic region, probably Africa, with subsequent spread of anatomically modern humans throughout the Old World. Workers in this school would toss the Neanderthals and other "archaic *Homo sapiens*" off the ancestral human path and even banish them to a species other than our own, for example, *Homo neanderthalensis*.[7] These nonsapient species supposedly became extinct during the last 100,000 years because of competition with true, anatomically modern *Homo sapiens*.

The molecular path back to an African or Asian mother Eve favors this last view: *Homo sapiens* is a few hundred thousand years old and originated in one geographic area. It too requires that the ancestral humans with the "correct" pattern of mitochondrial DNA left Africa or Asia, conquered the Old World, and in so doing outcompeted and supplanted contemporaneous populations of "archaic *Homo sapiens*" with the "wrong" mitochondrial DNA, such as Neanderthals. It also begs two questions: Where did all the populations of "archaic *Homo sapiens*" come from? And where is the fossil or archaeological evidence that *sapiens* and *sapiens*-like people commingled and competed?

Indeed, the remains of the different variants of fossil humans that lived in the Old World during the past 500,000 years are consistently recovered from different geographic areas; none, apparently, are found together at one locality. As with other fossil mammals, such a record implies that then, as now, *Homo sapiens* was a geographically diverse yet genetically single species of man.

This runoff between molecules and fossils is a replay of a paleontological debate that began two decades ago. The issue then was the timing of the ape-human split in evolutionary history. The molecules said 5 million years ago, the bones and teeth about 10 to 12 million, and neither side has changed its tune much since.

This, of course, awakens the old saw that history repeats itself. One would think, if history really repeated itself, learning history would be much easier. Apparently it isn't, because the lessons of mother Eve are not new. One lesson, judging from some of the press reports,[3] is that the mother Eve theory seems to appeal as much to noble ideology as novel science. The ideology says: Because mitochondrial DNA studies make the origin of man more recent, the family of man is more profound. Well, I would like to think that the species is just as much kindred, no matter whether *Homo sapiens* is geologically old or young, or whether molecules or fossils or creation tales track its ancestral road. Another old lesson is an irony of history: In creation tales, God does not alter the past, but in evolutionary ones, paleontologists and historians do.

# CHAPTER 8

Archaeopteryx *gliding in a late Jurassic forest in Bavaria, 140 million years ago: the oldest feathered bird or a dinosaur in borrowed plumage? (From Gerhard Heilmann,* The Origin of Birds, *D. Appleton and Co., 1927)*

# The *Archaeopteryx* "Hoax"

In the summer of 1860, workmen at a lithographic limestone quarry near Solnhofen, Bavaria, split open a slab of slate and exposed one of the greatest discoveries of the century: the exquisite imprint of a solitary feather. It was stunning proof of the existence of birds 140 million years ago, a time when dinosaurs were hitting their geologic prime, the period geologists call the Jurassic.

One year later the feather was overshadowed by a more remarkable specimen from a nearby quarry. Sandwiched between slab and counterslab lay most of the skeleton of this archaic bird, hollow-boned and fragile. It bore two unmistakably avian trademarks: a delicate wishbone (furcula) and rows of feathers imprinted along the wings and astride a long tail.

Paleontologist Hermann von Meyer named the bird *Archaeopteryx* ("ancient wing") *lithographica*. The specimen, owned by a local Bavarian physician, Dr. Carl Haberlein, was sold to the British Museum of Natural History in London along with other Solnhofen fossils for 700 English pounds. Haberlein had received the fossils from quarrymen as payment for medical services.

In evolutionary quarters, the London *Archaeopteryx* became a *cause célèbre*, an instant vindication of Charles Darwin's *Origin of Species*, which had appeared just two years earlier, in 1859. *Archaeopteryx* was the perfect "missing link" between birds and reptiles and dramatic fossil evidence for the continuous evolutionary thread between different kinds of animals. The wishbone and feathers made *Archaeopteryx* a bird. Otherwise, bone for bone *Archaeopteryx* would have passed for a small dinosaur, such as the crow-sized, bipedal (two-legged) coelurosaurian *Compsognathus*, skeletons of which were also found at Solnhofen.

Unlike modern birds, *Archaeopteryx* still retained teeth, a long tail, fingers at the ends of its wings, and ribs on the underside of its body—all anatomical ghosts of its reptilian heritage. Clearly, over 140 million years ago, a lineage of small, bipedal, dinosaurlike reptiles had acquired modified scales (feathers) for flight and/or insulation, as well as fused collarbones (a wishbone) for the attachment of flight muscles, and had crossed the threshold from "reptile" to "bird." *Archaeopteryx* was fossil proof of that event.

Sixteen years later, in 1877, the proof was reaffirmed. The feathers had barely settled over the first *Archaeopteryx*, when a second, more complete skeleton was unearthed in another Bavarian quarry, about ten miles from where the London specimen had surfaced. The skull was arched back over the neck in death's natural pose. Attached to the outspread wings were impressions of primary and secondary flight feathers, almost identical in detail to those of modern birds. Each vertebra of the long tail bore a pair of tail feathers.

Nationalism demanded that this *Archaeopteryx* stay in Bavaria, and in this case nationalism also ensured a high profit. This specimen, also owned by the Haberleins, had been purchased by Carl's son, Ernst, for 140 marks. They sold it for 20,000 marks to Werner Siemens, a benefactor who presented the fossil to the Humboldt Museum, Berlin. The main slab of the Berlin *Archaeopteryx*, with bones and feather impressions flawlessly preserved, has since become the most celebrated and published specimen of any animal, living or fossil.[1]

Well, fine feathers don't always make fine birds. At least according to Sir Fred Hoyle, the eminent British astronomer and one of the originators of the "steady state" theory of the universe. In 1985 he and his colleagues in England tried to kill two fossil birds with one dramatic accusation: The London and Berlin *Archaeopteryx* specimens were fakes that had duped paleontologists for over one hundred years. *Archaeopteryx*, they crowed, wasn't a bird in the hand, a missing link, or evolutionary testimony of the rocks, but a grand hoax.

Paleontological feathers aren't easily ruffled. But at the British Museum of Natural History it had been only some thirty years since they were badly clipped by the Piltdown Man affair.[2] (A human skull and an orangutan jaw—deliberately stained, filed down and planted in a Sussex gravel pit—had managed to pass for an extinct species of human ancestor (*Eoanthropus dawsoni*) at the British Museum from 1912 until the fraud was unmasked in 1954.)

Now there were allegations of more skeletons in the closet. The press, tasting "Son of Piltdown," gave Hoyle and his band of chicken-knockers more ink than Sir Fred ever got for his steady-state theory of the universe. The creationists, tasting that elusive triumph over evolution, declared *Archaeopteryx* a cock-and-bull story and Darwinism a dead duck.[3] One of the creationists was astrophysicist N. C. Wickramasinghe, a Hoyle colleague and co-accuser who had last been heard from at the 1981 Arkansas evolution trial when he testified on behalf of "creation science."[4]

The furor began when Hoyle & Co., using "advanced photographic techniques" on the London *Archaeopteryx*, concluded that the feather impressions were fabricated.[5] Their evidence of fossil quackery came from:

1. The "double-struck" appearance of the feather imprints, allegedly the result of a botched forging job rather than natural preservation
2. The poor fit of the main slab and counterslab of the London specimen, indicating that the forger had fooled with the fossil layers after the rock had been split open in the quarry
3. The finer-grained nature of the sediment bearing the feather impressions in comparison to the coarser sediment embedding the bones.

As Hoyle and colleagues reconstruct the conspiracy, the original fossil skeletons lacked feathers. The forgers—perhaps the Haberleins themselves—removed a thick layer of rock from around the bones, replaced it with a soft paste of fine-grained Solnhofen limestone and, using a real feather, imprinted the "fossil plumage" in the paste alongside the "wing" and tail bones. They also forged for a willing audience, providing the scientific world of the 1860s with what Darwinians had predicted—fossil evidence of intermediate forms and the evolution of one group of animals into another.

Another villain in the Hoyle plot is Richard Owen, an eminent paleontologist and anatomist, who was superintendent of the Natural History Department at the British Museum in 1862, and a virulent anti-Darwinian.[6] Hoyle postulates that Owen knew of (or worse, ordered) the fossil forgery and had the British Museum purchase the counterfeit specimen so he could tar and feather the Darwinians in their own phony evidence for evolution.

Trouble is, the case for fakery is just so much chicken scratch. Like evangelists, Hoyle and his flock have simplified issues and glorified goals. Their "advanced photographic techniques" amount to nothing more than low-angle flash lighting, fine-grained film and blowup prints. All of this may have been revolutionary in the Jurassic, but it was duck soup to Sir Gavin de Beer in 1954 when he published his exhaustive study—and superb photographs—of *Archaeopteryx*.[7]

Hoyle's three telltale signs of fossil shenanigans—"double-struck" feathers; poor fit of slab and counterslab; finer-grained sediment around the feathers—tell a tale that is explained by freshman paleontology.

1. Only Hoyle and colleagues are dumbstruck by the "double-struck" feather impressions. As de Beer and others have explained,[1] *Archaeopteryx* had *two* rows of slightly overlapping feathers on each wing, which would be preserved as overlapping natural impressions and appear double-struck. Also, in a quirk of preservation, the feather impressions on both the main slab and counterslab of the London specimen are of the *underside* of the wing. A logical forger would have faked one of the slabs with the upper side of the wing and the counterslab with the underside. Moreover, Hoyle and associates never explain how the forger managed to produce natural casts of the feather impressions on one slab and identical negative casts at exactly corresponding positions on the counterslab.

2. The poor fit of the main and counterslab of the London *Archaeopteryx* is the result of scientific, not surreptitious, tampering. Ever since the British Museum acquired the specimen, continued preparation of the slab (to expose more bones and feather impressions) and casting (to produce replicas for other museums) have altered the matching contours of the fossil-bearing surfaces.

It's also obvious that Hoyle, Wickramasinghe and friends haven't done much rock splitting. When stone slabs are cracked apart they fall from geologic grace: Bits of loose rock crumble away and, like Humpty Dumpty, the slabs can't be put back together again quite as perfectly.

3. The limestone around the feather impressions is indeed finer than that around the bones of the skeleton, but this comes as no surprise to geologists and paleontologists. The Solnhofen limestones, like most water-lain deposits, are composed of both fine- and coarser-grained layers. Fine structures, especially feather impressions, demand burial in fine-grained sediments to be preserved. If the grains composing the rock aren't fine enough, neither are the fossils. Also, algae, bacteria and fungi often coat the feathers and fur of decaying animal carcasses and trap much finer sediment than does the skeleton.

The rest of the "hoax" case also doesn't go according to Hoyle. First, in 1955, another partial skeleton of *Archaeopteryx* (called the Maxberg specimen) with faint but definite feather impressions was excavated near the site of the London specimen. Then, in 1973, a Solnhofen skeleton of a "juvenile *Compsognathus*" in Eichstatt's Jura Museum was reidentified as *Archaeopteryx*. It had been collected in 1950 and, as it lacked a preserved wishbone, had been mistaken for the small dinosaur until paleontologist F. X. Mayr noticed the weak outlines of wing and tail feathers. The obvious question is, did the forgers of 1861 and 1877 lay a plot that passed down through five generations and hatched ninety years later?

For that matter, if Hoyle is right, the fossil forgers must have been anatomical geniuses. Solnhofen yielded skeletons of small dinosaurs and *Archaeopteryx* that are virtually identical. Yet the forgers managed to doctor only those few small, dinosaurlike skeletons *with wishbones*, and (except for the Eichstatt specimen) avoided feathering the more common dinosaurlike skeletons *without wishbones*. There are expert paleontologists practicing today that aren't that good.

Just as perplexing is the arrangement of the "forged" tail plumage in *Archaeopteryx:* two feathers per tail bone—a pattern unknown in any other bird, fossil or living. The odds of the *Archaeopteryx* specimens in London and Berlin being the brainchild of fossil fakers are about as long as Foghorn Leghorn winning the Kentucky Derby.

In response, Hoyle and his fine-feathered friends dismiss the Maxberg and Eichstatt specimens of *Archaeopteryx* as small, featherless dinosaurs, although clear, close-up photographs display more plumage than Sally Rand or Gypsy Rose Lee ever wore.

Anatomy aside, the most damning blow to Hoyle's conspiracy theory is a faintly feathered skeleton of *Archaeopteryx* from Solnhofen that turned up in 1970 in a drawer in the Teyler Museum, the Netherlands. It had been collected in 1855, *four years before* the publication of the *Origin of Species*, and lay misidentified as an extinct flying reptile (a pterosaur) for over one hundred years. Other *Ar-*

*chaeopteryx* fragments were found even earlier, in the 1820s. Were these pre-1859 finds also faked? Did the fossil forgers presage Darwin's theory of evolution and predict the futures market in missing links? And all without benefit of *The Wall Street Journal.*

Cross-disciplinary science is wonderful. But when it comes to paleobiology, my advice to Hoyle is to quit chicken-necking, stick to star-gazing, and leave the bird-watching to others. Only five fragile feathered skeletons from Bavaria tell us of an *Ur*-bird, *Archaeopteryx,* 140 million years ago. They are not freaks of nature. They are thumping confirmation of an evolutionary continuity between "kinds," but not any more so than the bones of ancient three-toed horses, extinct "ape-men," or any of the countless fossils in museum cabinets and closets. Had *Archaeopteryx* never come to light, Darwinian evolution would still be safe, the descent of birds from reptiles still secure, and Sir Fred's record in paleontology still pristine. He should have known that if museums have skeletons in their closets, they're apt to be in the shape of whiskey bottles, not birds in borrowed plumage.

## Postscript

Just in case anyone might take the accusations of Hoyle and his colleagues seriously, paleontologists at the British Museum of Natural History undertook an exhaustive, yearlong study of the London skeleton. As expected, they demonstrated conclusively that *Archaeopteryx* is not a forgery.[8] The same verdict holds for the other four known specimens as well as the ones that will continue to be uncovered at Solnhofen and other quarries.

# CHAPTER 9

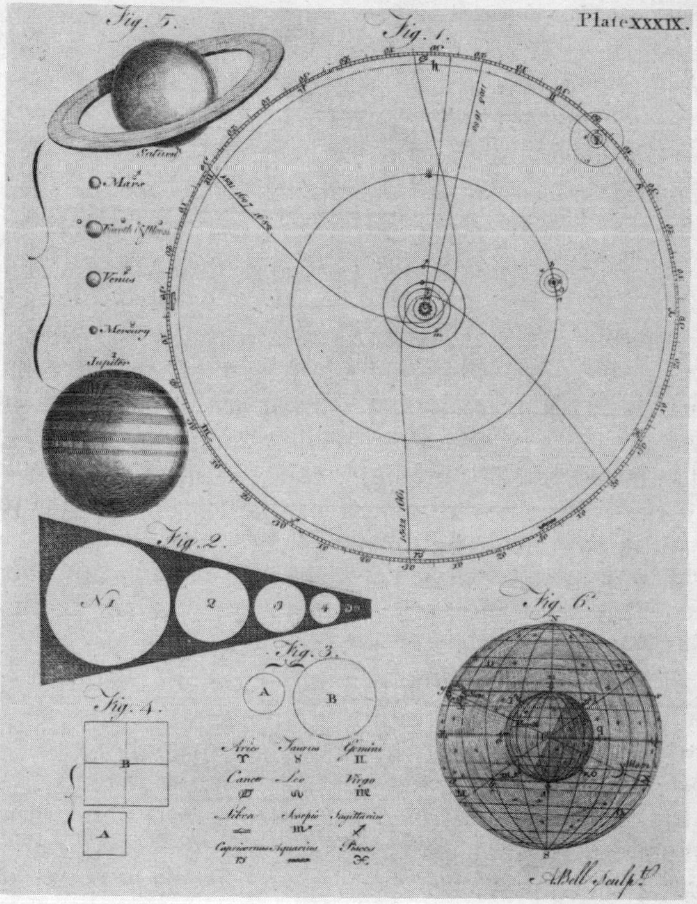


*The solar system as known in 1771 consisted of a sun, 6 planets (Mercury, Venus, Earth, Mars, Jupiter, Saturn), 10 moons and 21 comets. Concerning life beyond Earth, astronomers believed it inconceivable that "an infinitely wise Creator should dispose of all his animals and vegetables here, leaving the other planets bare and destitute of rational creatures." (From* Encyclopaedia Britannica, *1st edition, Edinburgh, 1771, p. 444)*

# Intelligent Life in the Universe

As a little boy, I often went for long winter walks with my father in the old neighborhood in Montreal. One clear cold night found us on Esplanade Street at the foot of Mount Royal, the city's mountain, crunching the hard-packed snow and breathing steam, my hand warm in his. I gazed up at the skyful of stars and asked my father one of the two questions parents dread most:

"How many are there?"

He studied the sky for a moment and answered:

"51,896,273,602."

He actually said, "Fifty-one billion, eight hundred and ninety-six million, two hundred and seventy-three thousand, six hundred and two." I was astonished. And just a little skeptical.

"Really?"

My father laughed. "Yes."

"How do you know?"

"Listen," he said, "if you don't believe me, count."

I may have been young and naïve, but I had enough smarts not to stand there ticking off a night sky of stars. When we got home I learned that my father was, at least, less of a fence-sitter than the encyclopedia. The book of knowledge intoned:

"There are many billions of stars in our galaxy."

Great. In other words, when you don't know, round off to the largely unknown numbers. Current astronomy texts are somewhat more politic. They round off to large whole numbers: "Our Milky Way Galaxy has about 200,000,000,000 stars." All those zeros add up to 200 billion, at which point the little boy in me wonders, "Who did the counting?"

Well, whoever did, the good news is that as of a few years ago there was one less star to count. It blew up in what astronomers call a supernova event. In the early hours of February 24, 1987, two amateur astronomers at the Las Campanas Observatory in the Andes of northern Chile spotted and photographed a brilliant flash of light in the Large Magellanic Cloud, an irregular dwarf galaxy that orbits the Milky Way about 170,000 light years from Earth and is one of the celestial landmarks of the Southern Hemisphere.

Actually, the star blew up about 170,000 years ago, when little Neanderthal boys were busy pestering their fathers. But the light emitted by the explosion just reached us in 1987. Reminds me of the Canadian Postal Service. Astronomers gathered excitedly at telescopes in observatories around the Southern Hemisphere to monitor the ancient light and witness a supernova firsthand. Only a hundred or so supernovas have been observed in recorded history, and only three in our celestial neighborhood—one by Chinese astronomers in A.D. 1054 (remnants of which we see today as an expanding cloud, the Crab Nebula), a second in 1572 by the Danish astronomer Tycho Brahe and a third by his one-time assistant Johannes Kepler, the German astronomer, in 1604.

Why all the fuss over a supernova? Its explosive death is the crucible for the ingredients of new stars, planets and life on those planets. Stars (or suns) burn hydrogen to produce the heavier element helium, and heat. In very hot, giant stars, the helium is burned and compacted into yet heavier elements—carbon, oxygen, magnesium, sodium, neon. Still hotter stars convert oxygen into sulfur, phosphorus and silicon. In stars twenty times the size of our sun and burning at a cool 2 billion degrees Celsius, iron, nickel and cobalt are welded from silicon. But the heaviest elements in the universe are forged by atomic fusion in the thermonuclear furnace of an exploding supernova—gold, silver, platinum, uranium, lead and copper. Even microscopic diamonds. The explosion hurls the elements out into space, eventually to be incorporated into new suns, new planets and, if all goes well, new life on those planets. The supernova in the Large Magellanic Cloud is a replay, 10 billion years later, of the fiery event that created the elements that ultimately made the Sun, Earth, rocks, life and man.

Astronomers quickly named the bursting star *Supernova Shelton* (or *1987A*), for Ian K. Shelton, one of its two discoverers. That's fine for Mr. Shelton, but for our Magellanic star it's kind of a backhanded, posthumous compliment. After all, the star is 170,000 years dead. The fiery image in our telescopes really isn't there anymore, only the fossil ghost of an event preserved by the relativity of space, time and light.

For that matter, astronomers may not be "seeing" *Shelton 1987A* at first hand or even at tenth hand. The light from the supernova explosion whizzed by many other solar systems and planets in our galaxy during its 170,000-year journey to Earth. Other intelligent, technological civilizations on those planets, if there are any, must also have picked up the light from the supernova explosion. So, *Shelton* may be old hat. The other civilizations may already have named it *Shmoishon* or *#pf1&?bzz@* or *Boom-Boom.*

All this raises a question larger than the number of stars out there: Is intelligent life orbiting one of them? To that my father answered a resounding "No." But astronomers and physicists answer with an equation. It's called the Drake equation, for Frank Drake, an astrophysicist who formulated the estimate based on earlier calculations by Enrico Fermi, Philip Morrison, Giuseppe Cocconi and other stellar scientists.

The Drake equation says it's simple to estimate the number of technological civilizations in our galaxy. Just multiply the following ten factors:

1. The number of stars in the galaxy (times)

2. The fraction of suitable galactic stars (times)

3. The fraction of suitable stars that can develop a stable planetary system (times)

4. The fraction of "stable planetary" stars that actually developed a planetary system (times)

5. The fraction of planets orbiting within the star's "zone of life"—not too close to or far from the star (times)

6. The fraction of zone-of-life planets with suitable atmospheric conditions for life (times)

7. The fraction of atmospherically suitable planets on which life has evolved (times)

8. The fraction of "living" planets on which intelligent species have evolved (times)

9. The fraction of intelligent species that have developed a technological culture (times)

10. The fraction of technological cultures still surviving.

Some simple equation. It's obvious we need intelligent life in the universe to calculate whether there's intelligent life in the universe. Another hurdle, of course, is defining intelligence. To judge by the intelligentsia, it's a way of thinking that allows you to enjoy something only until it becomes popular. Perhaps the best recent piece of evidence of intelligent life on Earth was a headline in the March 10, 1987, issue of one of the supermarket tabloids:

MAN WITH FOUR ARMS MAKES
FORTUNE AS DEODORANT TESTER

The Drake equation comes armed with some necessary assumptions. First, a technological civilization is defined as one that achieves the ability of interstellar radio transmission, allowing contact with other such cultures. Second, the data from planet Earth is our standard measure for events and processes on other planets in the galaxy. Astronomers call this the Principle of Mediocrity.

**Factor 1—The number of stars in our galaxy**—gives us a choice between the astronomy text's wishy-washy "about 200,000,000,000" and my father's no nonsense "51,896,273,602." Wishy-washy wins: 200 billion.

**Factor 2—Fraction of suitable stars**—eliminates the spent stars of the galaxy, those in the last twinkles of stellar life. After a few billion years of burning brightly, they began to run low on hydrogen fuel and, depending on their mass, either gassed outward into huge "red giants," shriveled to "white dwarfs" or collapsed in on themselves and exploded in a supernova. Whatever their fate, the death throes destroyed any planetary systems these stars may have had. In our galaxy, about 1 percent of the stars are red giants and 8 percent white dwarfs.

That leaves 91 percent of 200 billion, or 182 billion stars with a potential for having planets. Astronomers call these "main sequence stars" because they form an ordered sequence according to their temperature and luminosity—hot stars have high luminosity, cool stars low luminosity, and lukewarm stars medium luminosity. This principle was deduced independently in 1913 by the Swedish astronomer Ejnar Hertzsprung and the American Henry Norris Russell, and is now featured in every astronomy text as the Hertzsprung-Russell diagram.

**Factor 3—Main sequence stars that can develop a planetary system**—weeds out the weak, aged, short-lived and double stars in the galaxy, none of which have or are likely to have planets suitable for life. About half of the 182 billion main-sequence stars are the starry old-timers of the universe and probably lack planets. They condensed shortly after the formation of our galaxy, about 10 billion years ago, when the galaxy was still too young to have produced the raw elements for the formation of planets. Half of 182 billion leaves 91 billion stars.

But about half of these are double stars, a condition that induces very unstable planetary orbits. Most planets in a double-star solar system are quickly "deep spaced"—flung out of orbit after only a few spins around the star; remaining planets orbit too close to (too hot) or too far from (too cold) the double stars to maintain life. Half of 91 billion leaves 45.5 billion stars.

We also have to discard any main-sequence stars with lifetimes shorter than 2 billion years, the time it took photosynthetic plants on Earth to pump in and build up significant levels of free oxygen. The Earth is our only known living laboratory, and here oxygen was critical to life. We assume the same was true elsewhere. On Earth, oxygen fueled the ozone layer, the planet's protective shield against a star's harmful ultraviolet rays. As an energy source, oxygen powered the evolution of meiosis, cellular sexual reproduction, multicellular life, life on land, life in the big city, technological man and man on the moon. Only 23 percent of the 45.5 billion stars qualify for the 2-billion-year-plus lifetime for a similar oxygen buildup. That amounts to 10 billion stars that may have planets that may support life.

**Factor 4—Stars that actually have planets—**is an off-the-top-of-the-head guess. Astronomers think there's a fifty-fifty chance whether planets or thousands of small asteroids will coalesce around a star. Half of 10 billion leaves 5 billion suitable stars with planets.

**Factor 5—"Zone-of-life" planets—**tosses out those planets that orbit too far from or close to each of those 5 billion stars to support life. Life (certainly anything above microbial blobs) demands liquid water (not ice or vapor), and liquid water demands a star-to-planet distance that will allow a temperature range of 0 to 100 degrees Celsius on the planet. In our solar system, only 3 planets (Mars, Earth, Venus) orbit within this zone of life around our sun. But astronomers who fool with the Drake equation like to be conservative. They assume an average of 2 planets in the zone of life around each of the 5 billion suitable stars, which adds up to 10 billion zone-of-life planets in our galaxy.

Many of these planets, however, simply have the wrong mass and will violate **Factor 6—Zone-of-life planets with the right atmosphere for life.** Small planets, like Mars, have small gravitational fields, too small to keep enough of an atmosphere to maintain free oxygen or liquid water. Massive planets, like Jupiter, have such strong gravitational fields that they have retained their primordial poisonous atmosphere, composed of the most common galactic gases: hydrogen, methane, ammonia and water vapor. On Earth, the primordial atmosphere was replaced by the current one through photosynthesis and the outpouring of gases from the rocky innards of the planet. Even if the size of the planet is right, its atmospheric conditions may not be. Venus, about the same size as Earth, is a runaway hothouse with searing temperatures deadly to life.

Astronomers estimate that one fifth of the 10 billion zone-of-life planets in our galaxy may be the right size and have the right surface and atmospheric conditions. That's 2 billion planets ready for life.

And the probability of life on each of those 2 billion planets (**Factor 7—Planets on which life originated**) is 100 percent. If the conditions outlined thus far are met—right star, zone-of-life planet, right planetary size, right atmosphere—there is no reason why life should not evolve. It happened relatively quickly on Earth, about half a billion years after the crust stabilized and the oceans formed 4 billion years ago. The hot gases in the primordial atmosphere combined to form amino acids and nucleic acids, the first letters in the organic language; a diversity of cells and organisms followed. The 2 billion planets on which life could have formed were ripe for the same sorts of chemical processes.

The same 100 percent probability holds for **Factor 8—The odds of intelligent life** and **Factor 9—The odds of technology.** Once life evolves, intelligent life cannot be far ahead. And intelligent life, whales notwithstanding, breeds technology. Accordingly, there ought to be 2 billion industrialized planets out there. Except for one other complication: time.

On Earth, intelligent life and technology appeared 4.6 billion years after the origin of the planet, which is as good as any number to take as the minimum time required for an organism to evolve with the technological abilities of *Homo sapiens*. Our galaxy is about 10 billion years old, and stars and planets have been forming in it at a constant rate. That means, only about half of the 2 billion planets on which life evolved are old enough (5 billion years) to have allowed the evolution of a technological species.

That's potentially 1 billion species of intelligent life out there, each one busy trying to figure out whether there is any intelligent life out there. I hope the equations the other 999,999,999 species are using are easier to write home about.

**Factor 10—The survival factor.** Many of these 1 billion species in the galaxy may no longer be figuring anything. They may already be extinct, courtesy of their political or ecological shenanigans, or the death of their suns. Technological civilizations have a choice: They can survive for as long as they allow themselves to, or for as long as their suns allow them to. The technological clock starts ticking when the species learns to send radio communications over interstellar distances.

The crucial question here is "What is the typical or average lifetime of a technological species?" The answer is anybody's guess, so astronomers refer to it as *L* (for "lifetime"). *Homo sapiens* is working on an *L* of 69 years, if we start our technological clock in 1920 with the first commercial radio broadcast by station KDKA in Pittsburgh. Our potential lifetime on this planet is the same as the sun's. Like other stars on the main sequence, the sun can shine for 10 billion years. Five billion of those years have already gone into producing a four-armed deodorant tester. So, only 5 billion more remain for us to make contact with any of the other 1 billion technological species in the galaxy that are still alive.

How many of them are still alive? Las Vegas bookies

would place odds of $L$ to 5 billion on there being one technological civilization out there in the galaxy—that is, the ratio of the typical lifetime of a techno-species ($L$ years) to the potential lifetime (5 billion years). The odds of all 1 billion technological species in the galaxy still being alive are a billion times that (1 billion times $L$-to-5 billion), namely, $L$ to 5 or, in mathematical terms, the fraction $\frac{L}{5}$.

What kind of an answer is $\frac{L}{5}$? Is this the five-star finale to the Drake equation? It sounds more like a French shoe size, or a bingo number in Newfoundland. Clearly, astronomers need PR help in writing satisfying stellar equations. Equations should be short, energetic, substantial, yet relatively light. Something like $E = mc^2$.

Well, no matter how you write it, $\frac{L}{5}$ can't be solved unless we estimate $L$, the typical lifetime of a technological culture. Pessimists, using *Homo sapiens* as an example, estimate about 70 years, from the 1920 broadcast to a presumed nuclear holocaust by 1990. If the other billion civilizations followed this lamentable script, then there are only $\frac{70}{5}$ or 14 technological species left to talk to out there.

Optimists might say the $L$ is 5 million years, roughly the average geologic life span of many mammalian species. In that case, we have a chance of deodorizing 1 million technological cultures in our galaxy. Eternalists, of course, would vote for $L$ being 5 billion years, the remaining lifetime of the planet, which translates into a perspiring 1 billion advanced civilizations in our galaxy.

Vote for your own estimate of $L$ time. I'll be happy to serve as a pollster. In the meantime, I, for one, am glad we can't pinpoint the number of intelligent lives in the galaxy. It is one of those profound questions and, as my brother recently wrote me, "profound scientific questions must never be answered. Answers might be satisfactory. Satisfaction breeds sleep. And sleep would put an end to scientific inquiry."

# Chapter 10

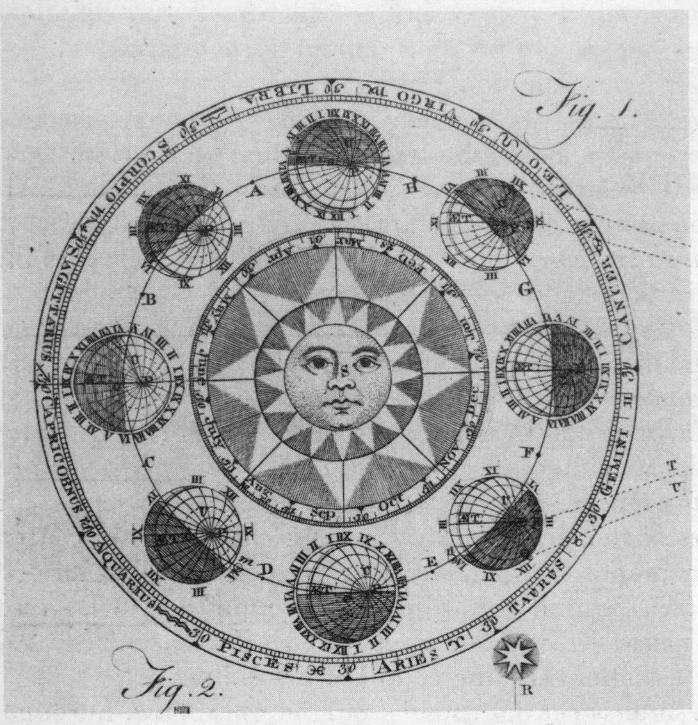


*A 1771 illustration of the revolutions of the earth about itself and the sun. No matter the date, the earth, moon and sun are far too out of step to have our calendar tell time with celestial grace. (From* Encyclopaedia Britannica, *1st edition, Edinburgh, 1771)*

# Gregorian Chance

Human time is the calendar, the landscape of passing time. Ever since man began to reckon by the sun and the moon and the earth, the calendar has been the crossroads for three passing times: sun time tells the year, moon time the month, and earth time the day. Trouble is, the times they tell are out of joint with one another. So the calendar became that magical mathematical wonderland where for a moment time could pause, recalibrate and harmonize the discordant orbits of sun, moon and earth.

The calendar of the Western Pennsylvania Conservancy is old fashioned. Unlike other calendars, it doesn't try to be the landscape of passing nudes, or Chevrolets, or boudoirs of the rich and famous. It is modest. Polite. Above each month hangs an airy watercolor of a wildflower of western Pennsylvania by Andrey Avinoff the artist. Avinoff the naturalist directed The Carnegie Museum from 1926 to 1945 after escaping from Bolshevik Russia.

A bittersweet nightshade, purple and yellow, introduces January 1988, which starts modestly enough with NEW YEAR'S DAY. January 4 presents a short, timely Gregorian chant printed politely in the space below 4:

Earth at perihelion (closest to the sun at 91.4 million miles).

There are almost 350 of these quick cantos of natural history, not quite one for each day of the year. January 5 quips, "A jackrabbit can reach speeds of nearly 45 mph." Speed is a recurrent theme. The 1986 and 1987 Conservancy calendars have already revealed that top speed for a cheetah is 70 mph, for a coyote 43 mph, for a giraffe 32 mph, for a sea turtle 20 mph and for a mole tunneling through the ground 15 feet per hour. Moreover, "As you read this you are traveling at 66,641 mph, the average speed of the earth's orbit of the sun."

The speed of the planet is breakneck, but the circuit is vast. The earth will take 365 days 5 hours 48 minutes 46 seconds to complete the solar tropical year, or the time to pass from the vernal equinox back to the vernal equinox.[1] The year is 11 minutes 14 seconds shy of mathematical convenience, namely, a flat 365 days 6 hours, or 365¼ days. In 46 B.C., Julius Caesar's chief astronomer, Sosigenes, mistimed the orbit by that 11 minutes 14 seconds when he pegged the Julian year at the convenient 365¼ days. The error was minute compared to what had passed for time in the early days of Rome.

March 4: "*March* is named for *Mars*, the Roman god of war." All roads may not lead to Rome any longer, but all Western calendars do. Rome's early calendar had 304 days spread over 10 months. March began the year, December finished it. The Conservancy reminds us that April came from the Latin *aperire*, "to blossom" (April 4/86), and May grew out of *Maia*, the roman goddess of growth (May 1/87). June, the fourth month, honored *Juno*, the "venerable, ox-eyed" wife of Jupiter and goddess of the Roman heavens. Quintilis was literally the fifth month, Sextilis the sixth. Similarly, September, October, November and December translated into the seventh, eighth, ninth and tenth months of the old Roman calendar (September 5/86). Clearly, sometime during the year, the Romans had 2 months, or 61 days, of missing time on their hands.

**January 7:** "January is named after the Roman god *Janus*, who had two faces and watched over doors, gates, and beginnings." The missing two months were formally added during the reign of Numa Pompilius, the second legendary king of Rome (715–672 B.C.). January was tacked to the beginning of the year and February to the end, and Romans abruptly switched from observing 10 months of Mediterranean Substandard Time to 12 months of Eastern Moonlight Time, the lunar year. The new calendar was based on 12 times the 29½ days (or 1 month) the moon takes to evolve through its phases. The 12 months alternated between 29 and 30 days for a total of 354 days in the year.

Numa's timing was bad on two counts. First, it too was mathematically inconvenient, as 354 was about 11 days too short to jibe with the solar year of 365. Second, 354 was an even number, then considered to be a bad omen. So they added one extra day to raise 354 to an odd 355, and they invented the leap year to deal with the conundrum that 355 was still 10 days and change short of an earth-round-the-sun year.

With the addition of the leap year, the calendar became a miasma, the crossroads of confused times. Numa ordered an extra month (alternately of 22 and 23 days) to be inserted every two years between February 23 and 24. The extra month, called *Mercedonius*, was supposed to help the Roman lunar calendar catch up to the solar year every 4 years. It didn't. Four times 355 plus a *Mercedonius* of 22 and 23 equals 1,645 days over 4 years or 366¼ days per year. The Roman calendar was running slow, time was losing a day a year and Rome was accumulating excess days.

**May 2:** "The earth weighs 5 sextillion, 887 quintillion, 613 quadrillion, 230 trillion tons." Time, especially surplus time, must have weighed heavily on the Roman minds. Their remedy, if one can call it that, involved a convoluted reshuffling of the leap year. In essence, the extra month, *Mercedonius,* was shortened to 22 days and inserted only three times in eight years (instead of the original four times), which saved 24 days over 24 years, at which point the whole bewildering shebang started over again. On top of that, the calendar shufflers also yanked February from the end of the year and put it after January.

All of this monkeying with time was to no avail. The calendar fell out of the hands of astronomers and into the regulatory clutches of the pontiffs and politicians. True to form, they introduced Career Saving Corrupt Time to prolong their stay in office. They fiddled with the extra month, enacted longer years and managed to delay annual elections. By 46 B.C. Roman time was horribly out of sync. January was falling in autumn, snow was falling in June and the spring equinox was falling in winter. The Roman calendar was living on borrowed time and, not surprisingly, so was the Roman Empire.

One obvious problem was that its days were numbered by donkeys and mooncalfs. Instead of counting days in sequence—July 1, 2, 3 and so forth—Romans counted backward from three fixed points of every month: the *calends,* the *nones,* and the infamous *ides.* The calends began the month, the ides marked the middle, and the nones fell nine days before the ides. So, March 18, for example, became the fifteenth day before the calends of April (April 1), March 9 was called the seventh day before the ides of March (March 15) and March 2 was referred to as the sixth day before the nones of March (March 7), which in turn was nine days before the ides of March (March 15), all counting inclusively. Whoa! Count it any way you like, it still adds up to the ides of the jackass. No wonder Cassius had that lean and hungry look. Neither he nor any other Roman (except some frazzled mathematician) knew what day it was.

It was left for Julius Caesar to rewind Roman time, although he never did anything about the calends, nones and ides of March or any other month. In 46 B.C., Caesar abolished the lunar year. The Julian year would be solar and, trusting Sosigenes, would consist of 365¼ days. The extra ¼ day each year would demand a leap year of 366 days every four years with the extra day legislated to fall between February 23 and 24. Seeing as how Rome and its spring equinox were by then about three months behind the times, Caesar made a one-time addendum of 90 days to the old 355-day lunar year, 67 between November and December and 23 at the end of February. As a result, 46 B.C., the catch-up year, lasted 445 days and was called the last year of confusion.

The first Julian year was 45 B.C. Its 12 months and 365 days were numbered into their present allotment. July (formerly Quintilis) was named for Julius (the month of his birth) and August (formerly Sextilis) for Augustus Caesar. August originally had 30 days, but seeing as Julius's month had 31, Augustus wasn't about to be slighted and short-monthed. He raided February for the extra day. New Year's Day was returned to January 1 and the spring equinox to March 25. Caesar had rendered unto human time and calendar an uneasy peace, splicing the dissonant pulls of sun, earth and moon.

**July 7:** "Earth at aphelion (farthest from the sun at 94.5 million miles)," halfway through its solar year. Unfortunately, Sosigenes's overestimation of the earth's orbit by 11 minutes, 14 seconds added an extra day every 128 years to the Julian calendar. To make matters worse, the Romans miscalculated again. They discovered Sosigenes's mistake, but figured it could be corrected by subtracting one day every 300 years. By A.D. 325 the March 25 equinox had marched ahead to March 21, and by the time A.D. 1582 rolled around, to March 11, much to the dismay of Pope Gregory XIII. After all, Easter Sunday in the ecclesiastical calendar was timed to the dependable recurrence of the spring equinox.

To right matters, the pope divined time: By fiat, he moved heaven and earth, banishing 10 earth revolutions from the history of the universe so that the equinox would occur on March 21 in the new Gregorian calendar. The 10-day theft actually occurred in October—by decree, October 4, 1582, was followed by October 15. That year the Gregorian calendar started to tell time in Rome, Spain, Portugal and France. Catholic Germany accepted Gregorian time the following year, but Protestant Germany, Denmark and Sweden kept Julian time until 1700.

Protestant England didn't go Gregorian until 1752, by which time the 10-day gap in planetary timekeeping had ticked into 11. The Gregorian calendar was adopted in Great Britain and the American colonies when Wednesday, September 2, 1752, officially ushered in Thursday September 14. People rioted in the streets. Who cared about the equinox? It wasn't worth the time of day if it meant getting short-changed by 11 days on a full month's rent. Russia didn't overthrow the Julian calendar until the October Revolution, which as a result took place in November and ought to be called the November Revolution. The Bolshevik coup d'état of October 24, 1917, Julian time, was bumped up 13 days by the newly instituted Gregorian calendar to November 6, 1917, the date the revolution is celebrated in the Soviet Union.

Most calendars now bide Gregorian time, including the Conservancy's. Exceptions include the Chinese, Hebrew and Muslim calendars. For example, the Hebrew calendar is lunisolar or semilunar, but rationalists would call it looney. Most years, called common years, have 12 lunar months, each with 29 or 30 days. An extra month is added 7 times in 19 years to make a "leap" year (officially, an embolismic year) of 13 months, so that Hebrew lunar time can leap up to solar time. Also, one day is added or subtracted each year to make the calendar fit the proper occurrence of holidays. Common years, then, can have either 353 days, known as a defective year, 354 days, considered a regular year, or 355 days, touted as a perfect year. Embolismic years are also defective (383 days), regular (384 days) or perfect (385 days).

A side effect of all these defective permutations is movable feasts and migrating holidays, the bane of the ecclesiastical and other calendars. Hanukkah, for example, fell on December 27 in 1986, on December 16 in 1987 and on December 4 in 1988. The Mayans too apparently didn't worry about drifting holy days. Their calendar year had 365 days flat, divided into eighteen 20-day weeks, plus a 5-day interval at the end of the year. Instead of tuning their calendar, as the Romans did, to account for the 5 hours 48 minutes 46 seconds missing from a true solar year, the Mayans moved their feasts each year.

Our week, like the Mayan one, has no celestial reason for being. The 7-day week mirrored the seven celestial bodies known in ancient times: the sun's day (Sunday), the moon's day, (Monday), Mars's day (Tuesday), Mercury's (Wednesday), Jupiter's (Thursday), Venus's (Friday) and Saturn's (Saturday). By modern astronomy, the week should be expanded to 10 days to honor Neptune, Uranus and Pluto. Perhaps a 24-day week is called for, given the 14 moons of Jupiter. The 7-day week is also mathematically at odds with the solar year (52.1428 weeks), the lunar year (50.571 weeks), 31-day months (4.428 weeks) or 30-day months (4.485 weeks), which means successive years cannot begin with the same day.

Perhaps the moral of human timekeeping is that our calendars have served their time and ought to be scrapped. The majority of countries may keep Gregorian time, but majority is no substitute for grace and art, and the calendar lacks both. It is an unnatural celestial hybrid, a horological mutant crossbred from three discrete species of time: the earth's motion about the sun (year), the moon's about the earth (month) and the earth's about itself (day). Classicists would shudder at the mythological monster: Saturn, the god of time, agglomerated from the bodies of Apollo, Diana and Gaia.

More to the point, calendars broadcast our state of knowledge. Every retinkered version of the calendar announced a new discovery in our measure of time and celestial mechanics. Thus, today's calendars are anachronisms, still broadcasting time according to the Middle Ages. After all, the Gregorian calendar marks time in Christian years: Bishop Ussher's 4,004 years B.C. (between Genesis and the birth of Christ) and 1,988 years *anno domini* (A.D.), or 6,012 years since biblical creation. The Hebrew calendar is counting through year 5748 since the same creation.

The timing of scriptural creation is not the issue. Rather, it is the simple proposition that the civil calendar ought to be in step with real time, namely, modern cosmology and our enlightened grasp of the age of earth and the universe. A second issue is, should our secular calendar still wear theistic clothes and run on reverent time? I don't recall the

First Amendment to the U.S. Constitution excusing the civil calendar from the separation of church and state.

Consider a modest proposal then: It's high time our civil calendar started marking the time that astronomy and geology tell. Despite its Gregorian backbone, the Conservancy calendar is conversant with geologic time: "300 million years ago most of W. Pa was covered by an inland sea" (December 19/87). "Amphibians and insects first appeared on earth about 325 million years ago" (May 18/86). "The cockroach is considered a living fossil, dating back nearly 300 million years" (May 11/88). "Sharks began to evolve over 300 million years ago; man has been around for only 2–3 million years" (March 7/86). And, perhaps most telling: "The sun is an estimated 4.6 billion years old" (March 23/87).

The new calendar, like the old, could start with solar time, 4.6 billion years, which is also the age of the earth and the moon. Of course, time really started ticking some 15 billion years ago, with the Big Bang origin of the universe, but 15 billion is too distant for meaning and too close to the national debt for comfort. Alternatively, we could start the calendar clock with the origin of primates 65 million years ago, or the first fossil record of *Homo sapiens* half a million years ago, or the beginning of the Holocene, the first epoch after the Ice Age, 10,000 years ago. Whatever the geologic starting date, the new calendar would be common ground for all human timekeeping.

Take 10,000 years. One solution is to forget weeks and months and think fractions, specifically, yearly fractions of 365 days 5 hours 48 minutes 46 seconds, or 365.24220 days. New Year's Day, 1988, would become .0027379 in the Year 10,000, and the next day, .0054758. The spring equinox would occur every year on .2241619 and June brides would wed between .416665 and .497260. Sounds romantic. Canada Day (July 1) would be celebrated on .5, Independence Day (July 4) on .5010952, and Labor Day (September 1) on .75. "Auld Lang Syne" would be sung again next year at the ringing in of 10,0001. No more moving feasts, or "thirty days hath September . . ." or leap years to remember.

A second, less fractious solution is to improve on the Mayan calendar. As before, start 1988 as the year 10,000. Institute eighteen 20-day weeks or, better yet, sixty 6-day weeks for an even 360 days in the year. Each week would consist of 4 work days and 2 off days, amounting to 240 days on the job, 120 off (versus the current 260 on, 104 off). Labor says yes, management is still thinking it over. Weeks could be given individual names, although#1–#60 is universal and easier to remember. All but one of the current days of the week could stay. I vote for zapping Wednesday. What about the 5 days 5 hours 48 minutes 46 seconds remaining in the solar year? Simple. Let's designate that "Huckster-week" or "Hustings-time" or "The Polympiad," place it after Week #60 and declare it the only time of the year that candidates for high office or low are permitted to campaign, confess or make political speeches.

# Chapter 11

XXXII.—Appearance of Man.

*Man in Europe in the late Pleistocene, about 35,000 years ago. Is one of these hunters "Ramtha," the Ice Age oracle with the New Age message: "Go west"? (From Louis Figuier,* The World Before the Deluge, *D. Appleton and Co., New York, 1867)*

# Backing into the Future

A friend of mine from rural Wyoming once wrote me that the West is where men are men and the plumbing is outside. Canadian National Railways first pulled me west from Montreal in March 1966. We chugged for three days through tree-choked boreal forests in Quebec and Ontario, across snow-buried prairie in Manitoba and Saskatchewan and into the foothills and mountains of Alberta. I got off in Jasper into a cold dusk and shivered down the main street past a Canadian Imperial Bank of Commerce. In front of the bank sat a handsome statue of a massively antlered deer on guard. When the statue snorted and the deer rose to greet me I hightailed it down the street, coat flying, suitcase bouncing. So much for men are men.

Obviously the West couldn't have been won by very much if the deer and the antelope were still playing in the street. A week later I met and mastered outdoor plumbing, but always with an eastern eye peeled for life-sized statues of deer. Clearly the West wasn't very safe, even more than a hundred years after Horace Greeley had published his "Go West . . ." recipe for success in the *New York Tribune*. No wonder. Greeley notwithstanding, the time-honored meaning of "go west" was to die, expire or, as they say in the West, go to grass with your teeth up.

The latest incarnation of westward ho comes from J. Z. Knight of Yelm, Washington, a small community fifty-five miles south of Seattle. Mrs. Knight refers to herself as a "channel"—a bodily conduit through which age-old reincarnated spirits speak to and teach those willing to listen about the meaning of life. Not surprisingly, one has to pay to listen, about $400 an earful.

Knight "channels" for a particular reincarnate she calls "Ramtha" who purportedly is 35,000 years old. His message, issued by Knight in throaty, rapid-fire clips for hours at a time, includes dire forecasts for the future and, among other bromides, the Pleistocene version of "go west." Ramtha warns of fire and ice, an apocalypse of earthquakes, volcanic eruptions, tidal waves, pollution and other natural cataclysms. The place to run, hide and stock a two-year emergency food supply is the Pacific Northwest, the safest of havens from nature's madness and human muck.

Apparently, hundreds have heeded this humbug, moving lock, stock and gullible to northern California, Washington and Oregon. Meanwhile, the Ramtha/Knight routine of channeling 35,000 years back to the future keeps channeling money back to the bank. It also attracted front-page coverage from the Sunday *New York Times* under the headline HUNDREDS ARE PULLED WESTWARD BY ANCIENT VOICE OF "RAMTHA."[1] In the article, Mrs. Knight "acknowledged that she was taking in millions of dollars a year from the fees collected at her personal appearances and from the sale of videotapes and other materials." A few Sundays later, ABC Television devoted five hours to Shirley MacLaine's experiences with the paranormal, including interviews with two of Mrs. Knight's competitors in "trance channeling."

The only spirits really worth investigating come in a bottle labeled single malt whiskey. Ramtha is no exception, not even if we were to feign a lobotomy and entertain this cockamamie notion of a reincarnated oracle. The few remaining synapses would still ask: Why should some yokel who lived 35,000 years back when have more insight into our future than, say, the premier of Saskatchewan?

Better yet, forget the future. Let Ramtha reveal the past. How can a paleontologist pass up Knight when she may be a true "missing link"? Anyone with a channel to 35,000 years ago qualifies as the Rosetta Stone of the Ice Age. Scientists would fight to fork over their $400 for Ramtha's instant replay of the Pleistocene scene—a mammoth hunt, or a Cro-Magnon conversation, or the latest in Neanderthal fashions. What about extinctions: Did more modern humans wipe out the Neanderthals? I'd give a case of malt whiskey to know whether Ice Age cave artists were portraying the local animals or just making a statement.

Ramtha should know, at least if he isn't a paleo-phony. Thirty-five thousand years ago he must have been an archaic *Homo sapiens*, much like a Cro-Magnon, who lived, perhaps painted, in the south of England during a warm break near the end of the Ice Age. It is a deduction that Sherlock would have called elementary. Ramtha's alleged 35,000-year antiquity places him squarely in the Old World, as humans didn't reach the Americas until 12,000 to 16,000 years ago. Modern man first appeared in Western Europe (in the guise of Cro-Magnons) about 35,000 years ago, suspiciously about the time when Neanderthals disappeared, and when the age of prehistoric art (cave painting, carving and engraving) began. Much of the Northern Hemisphere was frozen in glacial ice; the glacial episode (labeled *Würm* by geologists) had started 65,000 years earlier and wouldn't end for another 25,000 years. But during Ramtha's corporeal years, parts of Europe including southern England were ice free, courtesy of what geologists call an interstadial: a short period of minor glacial retreat. Ramtha's language is also a geographic giveaway;

his medium through Knight is an English learned, one imagines, in the land of John Bull.

Well, lobotomy or no lobotomy, Knight apparently hasn't asked Ramtha about the Pleistocene and Ramtha isn't saying anything except go west to northern California, Oregon, Washington and Idaho. The Pacific Northwest is grand terrain but, Pleistocene spirits aside, it isn't a refuge from earthquakes and volcanic cataclysms. Quite the opposite; the Pacific Northwest is a natural laboratory for seismologists and volcanologists, courtesy of plate tectonics, the nonstop conveyor-belt drifting and jostling of the twenty or so massive moving plates that form the hard outer layer of the earth's crust. Shouldn't we expect an all-knowing oracular spirit like Ramtha to be savvy about freshman geology, such things as geologic sutures, subducting plates, seismic gaps and suspect terranes?[2]

Take northern California. It is one of the notches in a belt of earthquakes and volcanoes that encircles the Pacific Ocean. The belt is called the Ring of Fire; it cinches the edge of the Pacific plate where that oceanic crustal mass meets, slides by, overrides or dives under neighboring plates. One neighbor is the North American plate. The Pacific plate slides inexorably north and west about two inches a year past the North American plate along a geological suture, the San Andreas fault. The suture is temporary: it holds during periods of quiescence and ruptures when the Pacific plate moves. Each rupture is an earthquake. Devastating ones occurred in 1906 (San Francisco), 1971 (San Fernando) and 1983 (Coalinga). Little ruptures are virtually continuous. During a typical forty-eight-hour period in December 1984, seismographs detected seventeen earthquakes along a portion of the San Andreas fault in central California, and twenty-five tremors farther inland near Bishop.[3]

Predicting earthquakes, especially ones of damaging force

(five or more on the Richter scale), is a geologic art that is the envy of spiritualists. Among the arsenal of geoprophetic measures is the "seismic gap," an area of active geologic suture (or plate boundary) where no significant movement has occurred for a long period of time since the last earthquake. Because adjacent portions of the plate boundary are moving, seismic gaps are under tremendous elastic strain to rupture and keep pace; they are earthquakes waiting to happen.

For example, Pallet Creek, fifty-five miles north of Los Angeles, along the Mojave section of the San Andreas, has averaged one large earthquake every 140 to 150 years during the past 1,400 years. The last one shook the Mojave in 1857, 130 years ago, and since then strain has accumulated at a rate of about 1½ inches a year, for a total of 16 feet of long overdue movement. According to the U.S. Geological Survey, there is a 50 percent chance that the Mojave is in for a good dose of seismic-gap relief—an earthquake of 8.3 on the Richter scale in the next 20 to 30 years.

The predictions farther north are no less earthshaking. The earthquake recurrence interval for the San Francisco area is a 7.5+ rocker on the Richter scale every 150 to 300 years. After 1906, the Bay Area remained seismically calm until 1956, but has since experienced a regular diet of moderate tremors—a grim repetition of events that preceded the 1906 quake. From seismic-gap measurements emerge the odds of a 6.5 to 7 Richter scale quake in northern California during the next 30 years: 40 percent to 60 percent for the area 60 miles south of San Francisco; 20 percent to 40 percent for the Bay Area itself; 10 percent to 20 percent for just south of Oakland; and about 5 percent for extreme northern California. The odds get better each year.

Oregon, Washington and Idaho are no strangers to geological tempests. From 30 million to 5 million years ago volcanic eruptions buried 50,000 square miles in these states under a mile of molten rock and formed a new landscape, the Columbia Plateau. The motor that drove this massive volcanism was also a plate to plate interaction called subduction. Along the western edge of the continent, ancient oceanic plates collided with and nose-dived under the North American plate, until the oceanic rock reached a depth of 60 to 100 miles in the earth's mantle. There, in a cauldron of terrific heat and pressure the rock became molten, rose in great plumes of hot lava, lifted the crust and punched through the earth's skin as volcanoes. This process feeds Krakatau, Mount Etna, Mount Fuji and thousands of less famous volcanoes near plate to plate subduction zones in Europe, Asia, Africa and ocean basins. Mount St. Helens and other active "plate-edge" volcanoes along the northwest coast bear witness that old oceanic plates are still being churned, recycled and readied for eruption under the western edge of North America.

Adding to the natural woes of the Pacific Northwest are hot spots, variants of volcanoes that are rooted deep in the earth's mantle below the plastic surface of a moving plate. When part of a continent (or ocean floor) drifts over an active hot spot, a pillar of pressurized molten rock forges its way up and through the crust to become a seamount, a volcanic ridge or a chain of volcanic islands—Hawaii, Iceland and Surtsey are good examples.

Hot spots can also be geologic blowtorches, heating, thinning, fissuring and rifting apart the continent that passes over it. A hundred and fifty million years ago South America was wrenched from Africa along the wake of the drift of that supercontinent across the St. Helena hot spot. Geologists have pinpointed two such hot spots under western

North America. One resides below Colorado. The other, under Yellowstone National Park, fuels the geysers and hot springs, fires the regional earthquakes and may be the site of future continental rifting along the Snake River plain.

Finally, Ramtha never tells his audience that much of northern California, Oregon and Washington is immigrant real estate. These states are a hodgepodge of what geologists call "suspect terranes,"[4] pieces of crust that have come crashing into the western edge of North America. The difference in spelling between "terrain" and "terrane" is intended, distinguishing between native and alien lands. Some Pacific terranes drifted from as far away as 2,400 miles and for as long as 150 million years before accreting to the geographic and geologic collage of the American west. For example, Wrangellia, a well-known terrane, plowed into Oregon 70 million years ago after a geological odyssey from south of the equator. It has since been splintered and faulted by geologic forces, its rubble strewn and welded into Vancouver Island, the Queen Charlotte Islands and the Wrangell Mountains in Alaska. There are conservative estimates of forty-six suspect terranes in the western United States; some geologists would say the figure is closer to two hundred. No matter the number, many of the inter-terrane boundaries are active or potentially active earthquake zones.

It's fitting that Knight lives in terrain as suspect as her spirits. Fitting too is the geologic irony of Ramtha-led throngs propelled like suspect terranes to the Pacific Northwest. Bad advice used to be cheap. Good advice, as my friend in Wyoming would say, is the kind you can chisel in granite. Start chiseling. If you're looking for mountains, fish and Bigfoot, go to the Pacific Northwest. If you're seeking seismic and volcanic calm, go to Saskatchewan.

# CHAPTER 12

*An 1886 rococo reconstruction of life in Wyoming in the late Eocene, about 45 million years ago. Front and center is* Uintatherium, *an extinct gigantic mammal here fancifully crowned with velvet antlers. (See Chapter 25, "*Scientific Literacy*.") Behind it looms an unknown, multihorned giant and facing it stands another nondescript creature. The other beasts in this Eocene zoo are: an extinct cat (crouched left); the saber-toothed* Machaeroides *(below the* Uintatherium *trunk); the three-toed horse of the time,* Orohippus *(foreground right); the primitive llama-looking camel,* Procamelus *(far right); a primeval pig (background right); and the forerunners of eagles (in the air), cranes (on the ground) and lemurs (in the trees, left). (From William Gunning,* Life History of our Planet, *Worthington Co., New York, 1886)*

# Bonehunter's Stew

The old bar in the Big Horn Hotel in Arminto, Wyoming, seats four and leans five. Arminto is a one-horse town on the eastern edge of the Wind River Basin. The odd cow-puncher or oilrigger wanders in now and then. October brings a mess of hunters gunning for deer, elk and antelope. Summers bring the bonehunters—the local tag for paleontologists. Carnegie bonehunters have been coming here for more than twenty years. But most days the bar is empty.

There was a time when the Big Horn saloon had to herd the patrons in and out, when the rooms on the second floor were crammed with sheepherders and ranchers. In 1914 Arminto boasted upwards of a thousand people. Arminto Station was the end of the line then for the Chicago, Burlington & Quincy Railroad. From here cars loaded with cattle and sheep headed east. Today the trains whistle past Arminto and stop elsewhere. A dozen or so ramshackle buildings are left, and about eleven people. The Big Horn Hotel now leans here and crumbles there, an old-timer staying extinction.

The bar is a long, languid sweep of blood-dark mahogany. Behind it rises the hutch, tall and straight, with ornate, tooled posts, dovetailed shelves, cabinets and a beveled mirror. The bar and the hutch were hand-carved in St. Louis a hundred years ago. The beer is cold when the chiller works, but often it doesn't. You can get a Bud, Miller or Olympia here, but you can't pump an Iron in Arminto, or much of anything else. The gas pumps across the street are glass, and they went dry long before Esso became Exxon. So did the wells. Drinking water is hauled in once a month.

The jukebox across from the bar is a chromed veteran. A quarter buys two memories from a fifties prom. Above it, a stuffed antelope head stares down on an acre of dance floor and a beat-up pool table. Another quarter buys 15 balls and a felt surface that has only slightly less topography than the Big Horn Mountains just outside the door. The antelope is a genetic oddity. Its right horn goes up, but its left one swoops down over its brow like a bony cowlick. Perhaps it was shot because it was the odd buck out. A door to the left leads to the dining room and kitchens. To the right is the front room and veranda of the hotel, which angles down to the southwest like a downwarped geologic stratum. A tired staircase leads to the guest rooms on the second floor. They are small, silent vignettes from the turn of the century: a brass bed, simple dresser and enameled washbasin.

The windows look out on sagebrush prairie running west to the Wind River Mountains. Rimming the rest of the basin are the Big Horn Mountains to the east, the Owl Creeks to the north, and the Granites and the Rattlesnake Hills to the south. The paved roads out of Arminto head for Waltman, Moneta, Hiland, Powder River, Hell's Half Acre and Shoshoni. The entire town of Hiland has been for sale for a few years. A dirt road runs west from Arminto, cuts through Cedar Gap, and winds along Badwater Creek through Badwater, Lysite and Lost Cabin. Here and there the miles of sage flats break into badlands, desolate canyons and mesas of purple, yellow, red and gray rocks. The breaks have names on topographic maps: Buck Spring, Deadman Butte, Dolis Hill, Wolton, Davis Draw, Alkali Creek. The rocks have names on geologic maps: Fort Union, Wind River and Wagonbed formations, and their subunits, Shotgun, Lysite and Lost Cabin members.

Names imply life, but the rocky badlands are still. Home on the range was never like this. The local ranchers—the Hendrys, the Spratts, the Sullivans—say that once in a while it rains bullfrogs in this part of the basin and you get a gully washer. Mostly, though, it's dry. Buck Spring or Wolton might grow a greasebush or two, but the badlands will sprout range grass when cows give beer.

The badland terrain is bad only above ground. Below the surface, the eroding rocks conceal the bones of ancient creatures that lived and died here over 50 million years ago. To bonehunters, this is fertile territory. Other paleontologists thought so too. From about 1880 to 1909 Jacob Wortman and Walter Granger mounted expeditions to this part of the Wind River Basin. For part of that time they were working for the American Museum of Natural History in New York. Old photographs show them in their field clothes: white shirts, black ties, vests and dark trousers. They beat us to the Big Horn bar and to the fossils in rocks nearby. But the bar wasn't in Arminto then. The Big Horn Hotel originally sat in Wolton, a small town seven miles to the west. In 1914, with Wolton going to dust, the hotel was ripped into two sections, hauled by horse teams to Arminto and reassembled. It's difficult to find Wolton now. A dot on the topographic map near the junction of the Okie Trail and Alkali Creek marks a few wooden skeletons.

Wortman and Granger discovered and worked the 50-million-year-old fossil lodes at Buck Spring and the other badland areas along Alkali Creek. All told, they collected the remains of about forty-five species of extinct mammals and reptiles. We've added many more since, especially the species of smaller mammals, some of which now carry *wortmani* and *grangeri* as specific names. In the back of our truck outside the hotel is the day's haul from one of our quarries at Buck Spring: chunks of rock with skulls, jaws, teeth and skeletons of animals long extinct.

Some of the jaws belong to a new species, an opossum-like marsupial, which we named *Armintodelphys*, after the town. We can put flesh and skin names on most of the fossil bones. Their size and shape are distinct and often can be matched to living evolutionary descendants or relatives. For example, many of the jaws in the rocks are minute, some no longer than a fingernail, and have teeth with tall, sharp, spirelike cusps and deep valleys. These jaws once chewed in the mouths of shrewlike mammals and primitive bats. There are also jaws with two large, curved chisellike teeth in the front that belonged to ancient, squirrellike rodents. One skull, about the size of a hazelnut, has huge eye sockets and a battery of square, low, rounded teeth. It is the first skull anyone has found of a particular ancient, tarsierlike primate called *Shoshonius;* Granger first described *Shoshonius* in 1910 from a fossil jawbone he recovered in the Wind River Basin, and named it after the nearby town of Shoshoni. The remains of five other primates occur in these rocks. A few of the jaws from Buck Spring are robust and bear long, dagger-shaped fangs (canine teeth); they belong to archaic hyenalike carnivores that preyed on *Shoshonius* and the larger animals from the quarries. The checklist of bones and beasts goes on: small, three-toed horses, tapirs and rhinos; lizards, crocodiles, garpike and snakes; a single frog with a pea-sized head. Finally, some of the bones belong to cat- and sheep-sized animals with gibberish names and no modern counterparts: multituberculates, condylarths, apatemyids.

It's difficult to conceive that 50 million years ago these wasteland breaks around Arminto were teeming with wildlife. Rocks are ciphers of their own history, the environment and forces that laid them down. Written in these badland rocks is a remarkably different era, an ancient, lush, tropical forest swarming with this strange fossil bestiary. Near Buck Spring, ribbons of yellow sandstone boulders wind through the midst of the badlands, past the quarries, following the course of the ancient river that meandered east from the Rockies. Acres of fine gray clay spread out from the yellow sandstone; they were deposited in massive sheets during frequent flooding of the river, each layer another flood. In the clays lie bits of petrified plant roots, wood and leaves, the preserved remnants of a huge, subtropical forest. Some of the clay layers are brick red, rusted by exposure to air for thousands of years when they were the ground surface and rich soil on which the forest grew and the animals trod. Some spots have thin, white, chalky rocks, formed in a series of semipermanent swamps and marshes on the floodplain of the river.

Fifty million years ago the area around Arminto was a luxuriant, dense rain forest seething with animals, something akin to parts of Central and South America today. Most of the small mammals and lizards lived on the forest floor and in the leafy canopy, feasting on the insects, fruit and vegetation. When they died, many fell or were washed into the swamps and marshes, buried in the fine mud and preserved. The quarry at Buck Spring is one of those ephemeral swamps. It and other localities nearby record the coexistence of about eighty species of mammals during a lone geologic moment 50 million years ago.

Today around Arminto you'd be lucky to spot more than three species of mammals roaming the range—mule deer, antelope and rabbits are fairly common. But the barren look of the land is no truer an indication of life here now than it is for life in the past. A recent wildlife survey of this part of the Wind River Basin turned up a whopping 75 mammalian species that call this habitat home. It's hard to believe, but here is the breakdown: 1 species of mole, 4 of shrews, 10 bats, 34 rodents, 6 rabbits, 3 canids (foxes and coyotes), 1 bear, 1 raccoon, 7 mustelids (weasels, ferrets, badgers and skunks), 2 cats, 4 cervids (elk, deer and moose), 1 sheep (bighorn) and the pronghorn antelope.

After 50 million years the bottom line of the organic ledger seems to balance. So much has changed, yet so little. Over the past 50 million years there was a slow environmental revolution here, from a green, wet, subtropical forest to a dry, windswept, scrub desert. Yet the terrain still supports about the same number of mammals. The diversity hasn't changed, just the species of organic players. Gone from the Arminto area is the multitude of ancient opossums, rhinos, tapirs, shrewlike beasts and *Shoshonius* and the other primates. Their evolutionary relatives and descendants survive today in the forests of Africa, Asia and South and Central America. Major new arrivals are the rabbits, shrews, bats, deer, antelope, the horde of rodents, and man.

This isn't much solace for Wolton, which is extinct, or for Arminto, which is walking the edge. Boom times may yet return to Arminto . . . well, minibooms anyway. I'm mindful of Yogi Berra's warning: "It ain't over till it's over." The Big Horn Hotel has been bought and sold three times in the past few years, and the new owners may cheat economic fate. One Sunday this July hundreds descended on the hotel for the 1st Annual Sheepherders Fair and Rodeo. That evening the band played some country, some western and some stomping blues. The old bar was again three deep in patrons and the dance floor a forest of boots. The front room, which hadn't felt such weight in decades, sagged a whisker more to the southwest. Earlier in the day our Bonehunter's Mutton Stew placed fourth in the Mutton Stew Cookoff. Never mind that there were only six entries. At the tasting the crowd raved about our stew, but two of the judges didn't. This may sound like sour mutton, but one of those judges was three sheets to a strong west wind, and the second couldn't taste a sheep from a goat. In any case, sheep are nothing more than wool's way of making more wool. The mutton is a sad by-product and bad aftertaste. At next year's fair we'll forgo the stew and enter the sheephooking event.

## Epilogue

The 2nd Annual Sheepherders Fair and Rodeo never made it to Arminto. The Big Horn Hotel and its long wooden bar became extinct when it burned to the ground in the early hours of Sunday, February 17, 1985.

# CHAPTER 13

LOXOLOPHODON CORNUTUS COPE.

*The fossil bones of this uintathere, a giant herbivore that roamed western North America 45 million years ago, were discovered in Wyoming in 1872 by paleontologist E. D. Cope. He immediately shot off a telegram to Philadelphia with the name and description of this new fossil species in order to beat his arch rival, O. C. Marsh, to press. Cope's version of the uintathere, although still somewhat clumsy, is an improvement on Gunning's reconstruction (Chapter 12) in skipping the velvet antlers and the elephant ears. (From E. D. Cope,* The Penn Monthly, 1873)

# Local Heroes

My boyhood hero was Jean Beliveau, the rangy, smooth-skating center for Les Canadiens de Montréal. I'd watch him play on *Hockey Night in Canada*, telecast religiously every Saturday night from the Montreal Forum by the CBC, Canada's state-owned network. Weekdays, I'd hear him on radio, stickhandling through the Bruins, Blackhawks, Rangers, Red Wings and the detested Maple Leafs, never breaking stride, the puck glued to his stick. Hockey at the Forum attained an artful grace with Number 4 roaming center ice.

Sadly though, no matter how often they praised him, play-by-play announcers in New York and Boston never learned to say Beliveau's name right: They called him "Gene Bellyview" instead of the French Canadian "Johhn Bell-ee-voh." The announcers managed not to massacre Bernie "Boom Boom" Geoffrion, another boyhood hero. Geoffrion, nicknamed for his booming slapshot, wore number 5 for the Flying Frenchmen. Maurice "The Rocket" Richard ("Reeshard") wore 9. Doug Harvey was 2. They were a dynasty. Molson Breweries owned them, Toe Blake coached them and the Canadiens ruled the ice in the arenas of the old six-team National Hockey League.

Eventually Beliveau retired, as was jersey number 4. Number 5 went on to pitch Lite beer commercials on television, the NHL went to twenty-one teams, Les Canadiens went mortal and my heroic notions left the Montreal Forum for the badlands of the American West and settled on bonehunters from the past: Jacob Wortman, William Diller Matthew, John Bell Hatcher and others. Unlike the hockey greats, they are local heroes: unsung, preserved on negatives rarely exposed, anonymous outside the arcane world of extinct beasts and bones.

Jacob Wortman first entered the Wind River Basin in Wyoming in 1880 to prospect for the fossilized bones of animals that lived there 50 million years ago. He found the dry, cracked, windswept surface of the badlands littered with the remains of forty-five species of extinct vertebrates (backboned animals: mammals, birds, reptiles, amphibians, fish) of which twenty-four were new to science. One hundred years later I arrived to prospect the same terrain and added about thirty more species to the census of the ancient bestiary.[1]

Wortman's name on old specimen labels speaks of classic collections of petrified skulls, jaws and skeletons he unearthed as he crisscrossed the basins and ranges of Oregon, Idaho, Kansas and Wyoming with horse teams and wagons, summer and winter, often under harrowing conditions. While prospecting in Oregon in 1877, Wortman and paleontologist Charles H. Sternberg meandered into the midst of the Bannock Indian war and had to flee on horseback for their lives, leaving their camp and belongings behind. In 1880, Wortman set out from Fort Washakie, Wyoming (near the present town of Lander), to investigate the Wind River and Bighorn basins. His horses, wagons and entire provisions were swept downstream in a ferocious current on his first attempt to ford the Wind River.

The following year was abnormally dry and Wortman and his party almost perished in the Wind River badlands from a constant shortage of water.

If perseverance is the judge, my name on specimen labels says less. One day in July a few years ago, while I was collecting in the Wind River Basin a century after Wortman, an icy rain and snowstorm took aim at our field camp. In fine pioneer fashion we abandoned tents, hopped in the truck and hightailed it eighty miles to town and the nearest motel.

In the 1870s and 1880s, Wortman was collecting for Edward Drinker Cope, a naturalist of independent means in Philadelphia and one of the giants of American paleontology. Cope was born in 1840 to a Quaker family and lost his mother when he was only three years old. By age six he was paleontologically precocious: After a visit to Philadelphia's Academy of Sciences, toddler Cope felt compelled to jot down his observations about the bones in the eye socket of a fossil ichthyosaur, an extinct marine fish-like reptile he had seen on display. Cope's teenage passion continued to be the natural history of amphibians and reptiles. After a stint at the University of Pennsylvania, he spent the Civil War years in Europe at various universities and museums, studying the biology and paleontology of backboned animals. Shortly after his return to the United States in 1864, Cope was appointed curator at the Academy of Natural Sciences in Philadelphia; from there his collecting expeditions to the American West became paleontological outposts of the U.S. Geological Survey.

The Survey's geological and geographical expeditions to the "Western Territories" were then exploring and mapping the lands west of the Mississippi River. Expedition naturalists took inventory of the life of the West, living and extinct. Plants and animals encountered on the way were collected, cataloged, crated and shipped back east; so were the remains of ancient beasts and fossil plants found eroding out of the gullies and scarps of the western badlands.

From 1872 to 1881 Cope organized, outfitted, led and sent paleontological expeditions to scour Wyoming, Nevada, Colorado, Kansas, Nebraska, South Dakota, Oregon, Washington, Montana, Idaho and New Mexico. In 1872 in western Kansas, Cope's expedition had to keep one eye out for fossils and the other for bands of Cheyenne Indians who were skirmishing with the local settlers. In the end, paleontology lost out to panic: One day, some of his own staff took off with the mules and provisions, abandoning Cope and the rest of his field party.

On a different expedition four years later, on June 25, 1876, Cope and his field crew passed near the junction of the Big Horn and Little Big Horn rivers while Custer and his regiment from the U.S. 7th Cavalry were being encircled and dispatched by Sitting Bull's Sioux forces. Again, Cope's guide and camp tender fled the scene, but the rest of the field party kept to their route through the Sioux territory, about a day's ride from Sitting Bull's camp on the Dry Creek of the Missouri River.

Cope's early interest in reptiles presaged his future role as one of the paleontological tsars in the dinosaur-collecting wars of the late 1800s. His note as a six-year-old on the ichthyosaur also presaged a prolific career. In 1859, at age eighteen, Cope presented his first scientific paper to the Philadelphia Academy, on the natural history of salamanders. He would write about fourteen hundred articles in all, mostly on North America's extinct groups of mammals and reptiles, and describe over six hundred new species of backboned animals.

One of those fourteen hundred publications is paleontology's most heroic printed legacy. We refer to it reverentially as "Cope's Bible," although Cope gave it the unassuming title of *The Vertebrata of the Tertiary Formations of the West: Book 1.* The tome is encyclopedic: It has 1,009 pages of text, 132 lithographic plates, is five inches thick and weighs as much as a small dinosaur (seventeen pounds). Cope's Bible describes and figures, in most cases for the first time, 349 species of extinct vertebrates that inhabited western North America during the first 40 million years of the Age of Mammals, from 65 million to 25 million years ago. Making up the Bible's 349 species are 234 fossil mammals, 62 reptiles, 52 fish and 1 species of bird—more than are enumerated in the King James version of prehistory. Book 2 of the Bible never appeared, but was slated to present the fossil backboned animals from the last 20 million years of the Age of Mammals. Needless to say, Cope's Bible remains the indispensable reference work in mammalian paleontology. When it was published in 1884 by the U.S. Geological Survey, it was free to qualified scientists. I paid a book dealer $35 for a copy back when I was a graduate student. A few years ago it still could be had for about $100 at antiquarian natural-history bookshops. Today Cope's Bible commands about $450 and its price is rising about 20 percent a year. Sell IBM, buy Cope.

As local heroes go, though, few rival John Bell Hatcher for grit, guts and an unfailing record of setting paleontology on its ear. Until recently, accounts of his life and expeditions were like the fossils he excavated: buried in scientific treatises and rarely read.[2] From 1896 to 1899 Hatcher led three expeditions to Patagonia, the inhospitable southern tip of Argentina. Each Patagonian expedition lasted a year, and each time Hatcher took only a field assistant, canvas tent, stove, wagon and collecting gear.[3]

By this time Hatcher was thirty-five and already legend: an ornery, Wild West loner and the finest bonehunter of his time.

Hatcher must have inherited some of his pluck from his mother, Margaret Columbus O'Neil. She was born in rural Illinois in 1842, and, while still a baby, was lost one winter day in a blinding blizzard. Miraculously, the baby survived long enough to be found by a family named O'Neil, who raised her as their own. At age fifteen Margaret married John Hatcher, Sr., and bore him five daughters and four sons; two other children were stillborn or died young. Hatcher junior was the second son, born October 11, 1861, in Cooperstown, Illinois. When he was ten or eleven, the family moved to a farm in Greene County, Iowa, near Lamoni. He received his early education from his father, who apparently awoke more of a learning spirit than he bargained for. Hatcher complained of farm life, remarking, "It's a poor family indeed that can't afford at least one gentleman," and went off to work in the local coal mines to earn the money for college.

Coal mining staked Hatcher to his tuition; the fossil plants in the coal beds kindled his interest in geology and the life of the past. He started his gentleman-bound journey at Grinnell College, Iowa, and ended it one thousand miles east at Yale College (the ancestor of Yale University), Connecticut. At Yale, Hatcher studied with James D. Dana, an eminent geologist and inspiring teacher, and came away consumed with the grand history written in the earth's rocks and a passion to collect and piece together the remains of its extinct life.

He was hired to do just that upon graduation from Yale in 1884. His boss was Othniel Charles Marsh, the doyen of nineteenth-century paleontology at Yale's Peabody Museum and an arch-rival of Cope. Marsh was born in Lockport, New York, in 1831 and, like Cope, lost his mother at age three. But their lives diverge beyond that coincidence. Although Marsh dabbled in paleontology as a teenager—he collected fossils from the diggings of the nearby Erie Canal—he subsequently showed more interest in acquiring a status in life than acquiring the remains of past life. Ironically, the two ambitions married in 1866, when Marsh's wealthy and philanthropic uncle, the importer and merchant George Peabody, endowed Yale College with a new museum and a position for Marsh as professor of paleontology. Uncle George had already financed four years of his nephew's education at Phillips Academy and six years at Yale.

Marsh's status rocketed as the expeditions he sent to the American West amassed vast fossil collections for the Peabody Museum, including some of the finest dinosaurs. He discovered the first pterodactyl (extinct flying reptile) remains from North America and published 270 papers, among them landmark monographs on dinosaurs, toothed birds, gigantic mammals called uintatheres and other groups of extinct beasts. Like Cope, for whom he had a bitter enmity, Marsh held important positions with the U.S. Geological Survey and a number of scientific societies. Unfortunately, the union of power and paleontological position made Marsh pompous in personality and imperious in professional opinion and behavior.[4] For example, in a vain attempt to outpublish Cope, Marsh virtually took over the *American Journal of Science*, a periodical founded and based at Yale, to ensure publication of his own articles as quickly as possible.

As Hatcher's employer, Marsh emulated that famous firm of eastern attorneys, Nasty, Cheap, Brutish & Short. Hatcher often went unpaid and was kept out west collecting fossils well into the winter months despite chronic debilitating bouts of inflammatory rheumatism. Marsh also kept Hatcher from studying and describing the superb collections of fossil animals Hatcher was amassing for Yale, reserving the scientific material for himself. Nevertheless, for nine years, from 1884 to 1893, Hatcher prospected and plumbed the badlands and eroded fossils of Nebraska, the Dakotas, Montana, Utah, Colorado and Wyoming. From there trainloads of skulls, skeletons, jaws and teeth of extinct mammals, dinosaurs and other beasts arrived in New Haven to stoke Marsh's fame and fuel his paleontological war with Cope.

Brewing in Hatcher's collections at the Peabody Museum were two paleontological revolutions. First, among the larger bones lay the grotesque skulls of strange, gigantic reptiles bearing horns, a beak, a massive shield of bone extending back over the neck and a battery of grinding teeth. One of these beasts later became a household word: *Triceratops*. What Hatcher had discovered was an entire group of extinct ruling reptiles, the Ceratopsia or horned dinosaurs.

Second, in 1889, in the same dinosaur beds in Wyoming that had eroded *Triceratops*, Hatcher discovered the first remains ever found of primitive, tiny, opossumlike fossil mammals that had lived alongside the dinosaurs in North America over 70 million years ago. Hatcher, ingeniously, had looked for the minute mammalian jaws and teeth on anthills, where ants dutifully deposit any small pebble—and fossil—encountered in their subterranean tunneling. He immediately exploited his newfound army of unwitting bonehunters by planting additional colonies of ants throughout the badlands of the American West for future harvests of tiny fossil teeth and bones. Ever since Hatcher, paleontologists have robbed anthills.

Revolutions and ants notwithstanding, government support for paleontological research dwindled. In January 1893, Marsh lost his funding from the U.S. Geological Survey. Hatcher, in the third year of a five-year contract, was out of a job. But not for long. He was hired by William Berryman Scott, the paleontological luminary at Princeton University, and immediately sent back out west to do for Princeton what he had done for Yale.

It was fortunate that Hatcher preferred field collecting to working in the laboratory, because he marched to a different paleontological drum than his co-workers. Hatcher had a single-minded, hard-nosed, perfectionist temperament that ran headlong into colleagues and employers like Marsh and Scott. Field expeditions slaked his hankering for the great open spaces where his cantankerous, uncompromising tenacity went unchallenged. He worked alone or, at most, with a few students and assistants. This attitude culminated in a wildly ambitious scheme that Hatcher hatched in 1894 or 1895 while collecting in the West for Princeton: He would explore and mine the fossil-rich rocks of Patagonia.

Like Humboldt and Darwin before him, Hatcher was one of a succession of illustrious naturalists lured to South America by its exotic faunas, living and fossil. What intrigued Hatcher in particular were the strange, extinct marsupial (pouched) and placental (nonpouched) mammals preserved in Patagonia's lava-strewn badlands and towering sea cliffs: marsupial sabertooths, hyenalike beasts and small, ratlike creatures; and alien herds of large, placental, hoofed, grazing animals that superficially resembled rhinos and horses, but had short trunks protruding in front. This strange extinct bestiary had been discovered and heralded by the Ameghino brothers, Carlos and Florentino, paleontologists and local heroes in Argentina. The animals were radically different from any Hatcher or anyone else had found in North America. No wonder. For some 80 million years South America had been a drifting evolutionary experiment, its animals and plants arising, evolving and dying in isolation on an island continent.

The first Hatcher/Princeton expedition sailed from New York for Buenos Aires in February 1896. With Hatcher was his sole assistant, O. A. Peterson (his brother-in-law and a fellow paleontologist at Princeton), his field equipment and the little money he had managed to raise from private sources to finance the trip. In the end, most of the expenses for Hatcher's three Patagonian expeditions came out of his own, rather destitute, pocket. It is rumored that Hatcher, a poker shark, taught the moneyed Patagonians how to play the game, cleaned out their wallets and spent the winnings on the expeditions. He also entered Argentina under a hastily forged medical certificate, and eventually smuggled caches of fossils out of Patagonia via Chile to circumvent Argentina's strict antiquities laws.

By June 1896, after a month in the field, Hatcher and Peterson had collected one and a half tons of fossil skulls and skeletons of the mammals that had roamed Patagonia some 20 million years earlier, during a geologic epoch called the Miocene. At a second locality, the foot of a sea cliff called Corriguen Aike, they amassed four and a half tons of Miocene fossil mammals during September and October of that year. Here, collecting was a race with the sea. Hatcher and Peterson would follow the receding tide, chiseling out the skulls and skeletons preserved in the smooth beach rock that moments earlier had been under sixty feet of churning ocean. When the tide turned, they raced ahead of the oncoming sea, picking up the excavated specimens as they ran and carrying them to safety beyond the reach of high tide.

During the three expeditions, Hatcher also trapped, shot,

skinned and preserved great numbers of the living mammals, birds and reptiles of Patagonia for zoological museums in the United States. He fed himself and his horses from the land and survived the fierce blizzards of the Patagonian winters huddled around the camp stove in the canvas tent. Twice he came close to death: once during the first expedition when a freak accident with his horse opened his scalp to virulent infection, and again during the winter of the second expedition, from a paralyzing, six-week attack of inflammatory rheumatism.

Throughout, Hatcher continued to traverse Patagonia in search of fossil-bearing rocks. He became seduced by the grand contours of its coastal, interior and Andean landscape, much of which he was the first to explore and map. He discovered and named new mountains, rivers and lakes and, at one point, his geographic explorations made him the hero of diplomatic circles in Washington. Hatcher had defused imminent war between Argentina and Chile over their southern boundary by demonstrating the irregular course of the continental divide—and the Argentinean-Chilean border—in the Patagonian Andes.

In 1899, following the last expedition, Hatcher feuded with Scott, as he had earlier with Marsh. Hatcher put it bluntly in his Patagonian field notes: Scott, like Marsh, was a "fireside naturalist who seldom goes beyond his private study"; both were "parasites" who rode his paleontological prowess to fame and denied him the opportunity to study and publish either his massive collections of fossils or the geology of the regions that yielded them.

By 1900, Hatcher and Peterson had left Princeton and come to Pittsburgh to work for Andrew Carnegie and William J. Holland at The Carnegie Museum. Here Hatcher, although racked by rheumatism, began to organize a fourth expedition to Patagonia and Antarctica. He wanted to test the radical notion that the southern continents had at one time formed a single landmass. It was an idea that would be published a decade later by Alfred Wegener as continental drift and documented almost fifty years later by geologists and paleobiologists as plate-tectonics theory. Gondwana would become the name given to that ancient, single, southern landmass of Africa, Australia, South America and Antarctica. Hatcher's plans were dashed by a lack of funds and his untimely, almost ironic, death.

His end, like that of most heroes, does not seem fitting. On July 3, 1904, five years after surviving the relentless harshness of the Patagonian expeditions, he succumbed to typhoid fever contracted from Pittsburgh's untreated drinking water. On July 6 he was buried in an unmarked grave in Homewood Cemetery in Pittsburgh, alongside his daughter, Ruth, who, six months earlier, at age two, had died of scarlet fever. Also not fitting was the fate of his massive collections of Patagonian fossils. For eighty-six years they had been cataloged, stored, studied and displayed in the paleontological rooms of Guyot Hall, Princeton University. But, in 1985, the Geology Department at Princeton decided to forsake almost a century of tradition, close its paleontological shop and give its fossil collections to another institution. The ghosts of Marsh and Scott conferred and the cabinets with the skulls and skeletons of ancient beasts, including Hatcher's hard-won lode of Patagonian fossils, were packed, crated and shipped down the turnpike from Princeton to Yale.

# CHAPTER 14

Daubentonia, *the aye-aye of Madagascar, was named for Jean-Marie Daubenton, a naturalist who managed to save his neck from the guillotine during the French Revolution. (From Daniel G. Elliot,* Monographs of the American Museum of Natural History, *1912)*

# Measure for Measure

Revolutions are akin to fishing. Sometimes they catch fish, other times they just drown worms. Essentially, revolutions tend to bring a sudden change in misgovernment. A good example is the French Revolution, which caught a few fish at the expense of a lot of drowned worms. It started in 1789, a decadc after the American Revolution. The French imported the American model but replaced Yankee Doodle with the swoosh of Dr. Guillotin's invention.

The revolution gave France a new hero in Napoleon Bonaparte, especially those Frenchmen trapped in a five-foot-two body. Unfortunately, Napoleon also became the model for the little dictator, a new species of tyrant that quickly flourishcd in tropical, temperate and banana republics. Their notion of "equality and fraternity" meant equal oppression for the short and the tall. For example, after the Napoleonic wars ended in 1815, a new French census discovered that the average height of Frenchmen had dropped by two inches. This is one of the unnatural selections caused by armies, war and revolutions.

The French Revolution is applauded for dethroning the aristocracy. Fine. But it turned right around and imposed a new tyrannical rule on the world—the metric system.

We are stuck with the metric system. Any scientist who dares measure anything measures it in metric, no matter if it's the size of a slug or the mating territory of a moose. Paleontologists are no exception. Fossil teeth, jaws, skulls, skeletons and shells are duly measured in millimeters; geologic strata, in meters. Citizens of metric countries are obviously also stuck with the metric system. They weigh in grams, grow by meters, drink in liters, freeze in Celsius, farm in hectares, and live under kilopascals of barometric pressure.

Scientists in nonmetric countries, such as the United States, are like Jekyll and Hyde, living under a schizophrenic rule. They have two yardsticks. Day-to-day life comes packaged in pounds, feet and Fahrenheit, whereas the measure of their scientific work is milli-this and kilo-that. One would think that the metric fiat comes with the halo of precision. Not so.

The metric system was the brainchild of one Gabriel Mouton, vicar of Lyons, in 1670. Its time, however, came more than a hundred years later, in 1790, in the midst of the French Revolution. The Revolutionary Convention decided to purge the old system of weights and measures along with the old monarchy, and proposed a new standard, the meter. It was to be one ten-millionth of an earth quadrant, namely, one ten-millionth of the distance from the equator to the North Pole measured along the land meridian that runs through Paris. Some standard. It's one thing to propose the meter. It's quite another to go measure the length of a standard quadrant of an earth meridian. But, then again, the distinction between ideas and ideology was never a forte of revolutionaries.

After much debate, the Revolutionary Convention gave up on the quadrant and settled for the distance from Dunkirk to Barcelona, which is one ninth of the quadrant. The meter would then be one ten-millionth of nine times that distance. It took seven years for the French to pace off the footage from Dunkirk to Barcelona, multiply by nine, divide by 10 million, and arrive at the standard length of the meter.

As it turns out, they undershot the measure by about two parts in ten thousand, or .00022883 meters. This is probably no more accurate than the old standard for the Scottish inch—the average width of three men's thumbs—or the German foot—the average length of the shoes of sixteen men leaving church one Sunday morning in the seventeenth century.

Accurate or not, the meter determined the liter, the metric measure of volume (one tenth of a meter cubed), which in turn defined the kilogram, the metric unit of weight (the weight of one liter of water at 4 degrees Celsius). Area was measured off in ares (100 square meters) and hectares (100 ares or 10,000 square meters).

Metric measure became law in France on December 10, 1799, and its use compulsory on January 1, 1840. By 1911 the metric system had accomplished what Napoleon could never do. It had invaded and conquered Argentina, Austria, Belgium, Brazil, Chile, Germany, Greece, Hungary, Italy, Mexico, the Netherlands, Norway, Peru, Portugal, Romania, Serbia, Spain, Sweden and Switzerland. Its use became legal but not obligatory in Egypt, Great Britain, Japan, Russia, Turkey and the United States.

Canada, ever in search of a national identity, went metric a few years ago. After hundreds of millions of dollars spent on advertising and retooling, there was no measurable improvement in Canada's trade, balance of payments or identity. The yardstick of political honesty didn't increase the extra few centimeters it takes to convert the yard to the meter. Canadian weather felt colder in Celsius. Metrification merely provided a legitimate excuse to jack up the price of gas, beer and milk, and to change the weight of dinosaurs displayed in Canadian museums from tons to tonnes.

Today, only the United States, Yemen, Brunei, Burma and Liberia remain holdouts to the legislated use of the metric system, an alliance surely as worthy of recognition as NATO and OPEC. "USYBBL" would be a fitting acronym. At least these countries have avoided the trauma and cost of a metrification transplant. And they've heeded the Golden nonmetric Rule: If it ain't broke, don't fix it.

Had the English-speaking world been a metric one, we would be heirs to a cultural wasteland. Pound for pound (kilogram for kilogram?), metric units are as dry as dust. For the bards, avoirdupois was the universal measure of prose and poetry. For example, Shakespeare's Shylock never demanded his 454 grams of flesh, and Robert Frost never had "kilometers to go before I sleep." How far would Erskine Caldwell have gone with *God's Little Hectare*, or Jules Verne with *96,000 Kilometers Under the Sea*? Clearly, in literature an ounce of metric prevention was worth a pound of cure.

Nor were songwriters atuned to Napoleon or the metric system when they wrote "Five foot Two, Eyes of Blue." I doubt if the metric lyrics—"Fifteen Point Seven Decimeters, Eyes of Blue"—would have kootchy-cooed into the top-forty chart. Neither would the metric version of "Sixteen Tons and What Do You Get." You get 14.51 metric tonnes and Tennessee Ernie Ford wishing he were 1.82 meters under. Metric units have made one inroad into day-to-day language: Kilo is the measure and talk of the drug trade.

To be honest, the metric system did bring a measure of uniformity to European weights and measures. In the pre-guillotine days before the French Revolution, the natural and commercial world was measured in a bewildering array of units: *milles* (1.2 miles), *marcs* (roughly half a pound), *ells* (27 inches), *arpents* (about 1¼ acres), *brasses* (5.3 feet), and *perches* (22 square feet) are just a few. Italians had over 200 different measures for the "foot," a quibbling over small distances that seems to arise in small countries.

The eminent French naturalist the Comte de Buffon, who was keeper of the Jardin du Roi (Royal Garden) and the Cabinet du Roi (Royal Museum) for kings Louis XV and Louis XVI, used *pieds* (the French foot of 12¾ inches), *pouces* (one twelfth of a *pied*, or 1.02 inches) and *lignes* (one twelfth of a *pouce*) to measure and describe the organisms in his multivolume animal encyclopaedia *Histoire Naturelle*. Buffon died in 1788, a year before the Revolution and the beginning of metrification.

Antoine Lavoisier, the discoverer of oxygen and one of France's greatest chemists and agriculturalists, was not so lucky. As a member of the French Academy of Sciences, he was on the commission appointed by the Revolutionary Convention to develop and implement the new metric system. But Lavoisier's scientific accomplishments and agricultural programs ran headlong into the anti-intellectual fervor of the Reign of Terror and he was guillotined on May 8, 1794, at the Place de la Révolution.

Two months later, Buffon's son was also beheaded. At the time he was being tutored in zoology and botany by Jean Baptiste de Lamarck, a colleague of Buffon the elder and one of the early contributors to evolutionary thought. Lamarck cheated the guillotine through political shrewdness and courage. In 1792, he faced down a rioting mob intent on trampling the Jardin du Roi. Later, he persuaded the Revolutionary Convention to nationalize and preserve the Royal Gardens and Royal Museum as France's

National Museum of Natural History. Lamarck survived, but he is less remembered for his evolutionary opinions or for saving the royal scientific collections than for his flawed theory of acquired inheritance in plants and animals.

A much better reputation follows Louis Jean Marie Daubenton, who also ended owing his life to natural history. He too studied at the Royal Gardens, and he had provided the anatomical dissections and descriptions of 182 species of mammals and reptiles for the first two volumes of Buffon's *Histoire Naturelle.* But the two had a falling out. Buffon lost Daubenton's expert assistance, and, as a result, the subsequent volumes of the *Histoire Naturelle* lost precision and scientific prestige. Daubenton became a lecturer in natural history, rural economy and mineralogy, and devoted his studies to comparative anatomy of animals, paleontology and agriculture. What saved his life was his experiments in the husbandry of merino sheep. Daubenton was spared by the Revolutionary Convention after he demonstrated ways of increasing wool production in postrevolutionary France. Today, the primate *Daubentonia*, the aye-aye of Madagascar, carries his name.

At the same time the French Revolutionary Convention was executing naturalists and fiddling with weights and measures, it also went fishing for a new calendar. September 22, 1792, was to be the first day of the first year of the new republic. The Gregorian calendar was given the heave-ho and the twelve months of the new calendar were renamed and reordered by season into rhyming groups of three:

Autumn *Vendemiaire* (the vintage:
September 22 to October 21)
*Brumaire* (the foggy)
*Frimaire* (the frosty)

| | |
|---|---|
| Winter | *Nivose* (the snowy) |
| | *Pluvose* (the rainy) |
| | *Ventose* (the windy) |
| Spring | *Germinal* (the blooming) |
| | *Floreal* (the flowering) |
| | *Prairial* (the meadowing) |
| Summer | *Messidor* (the reaping) |
| | *Thermidor* (the heating) |
| | *Fructidor* (the fruiting) |

British reaction to this poetic nausea was swift. They lampooned the new French calendar with their own rhyming triplets: Autumn—Wheezy, Sneezy and Freezy: Winter—Slippy, Drippy and Nippy; Spring—Showery, Flowery and Bowery; Summer—Wheaty, Heaty and Sweety.

Fortunately, reason prevailed over revolution and the new French calendar went the way of drowned worms. In case you're wondering, had the new Republican calendar followed in the conquering wake of the metric system, the date of this year's World Series would be Vendemiaire, Year 197.

# CHAPTER 15

Bos, *the ancestor of designer cows, was domesticated and bred by man to be long on bulk and short on smarts. (From* Encyclopaedia Britannica, *1st edition, 1771)*

# 'Til the Cows Come Home

A few years ago, while heading west on Interstate 80, I discovered two little-known laws of nature. The lesser law is that there's not much else to do across Ohio and Illinois except discover laws of nature. Anyone who has ever driven I-80 knows that crossing Ohio is about as exciting as kissing your sister. The output of natural laws would have tripled had Darwin, Newton or Einstein been able to take The Carnegie Museum's Chevy Suburban down the stretch of highway between Toledo, Ohio, and Peru, Illinois.

The greater law concerns cows—at any one time, all cows in a group in an open field tend to facc in onc dircction. The tendency is strong through Pennsylvania, Ohio, Indiana and Illinois and becomes overwhelming for cows in Kansas, Iowa, Nebraska and Wyoming. I haven't tested the law on cows farther west only because our three-day haul across the country every June ends in the fossil-bearing badlands of the Wind River Basin in central Wyoming, near the small town of Lost Cabin. But I can swear by Texas cows from the paleontological time I've done in that state. Ditto for New Mexico and Colorado cows. And I'll vouch for foreign cattle north of the border in Alberta, from five highway-years of cow-watching between Edmonton and the Red Deer River badlands near Brooks—badlands that have eroded many of the finest specimens of Cretaceous dinosaurs that grace natural-history museums.

I won't vouch for California cows though, where natural laws may not apply. There, instead, many of the humans tend to face in the same direction, which most recently was toward Mount Shasta, during a "New Age" happening. Mount Shasta is apparently a mystical piece of geology: When plied with enough money, trance-channelers induce it to issue ancient vapors of "psychic energy" upon the gathered multitude. For a few more bucks they'll make a cow jump over the moon.

Clearly, the Cow Law isn't just another piece of sensationalist bulldocia intended for a week of screaming headlines in the *National Enquirer.* Like all other great laws of science, it should receive the imprimatur of the National Academy of Sciences, not to mention *The Old Farmer's Almanac.* After all, the fact that cows in a herd in an open field face in the same direction is empirically testable. And testing is what I've been doing every summer on I-80 ever since the cow hypothesis became law, much to the bawling and beefing of my colleagues in the truck. They're sick and tired of my triumphant gesticulations every time we pass a group of Black Angus or Charolais aligned in a field like used Chevies on a car lot. I'm not surprised. After a while, Einstein's colleagues too were probably fit to be cowtied at the mere mention of "relativity."

Why should a group of cows in an open field be oriented in one direction? Maybe they were facing the Dairy Queen down the road. Or the bullpen in Yankee Stadium. Or Ottawa and their relatives in Parliament. Or the *plaza de toros* in Madrid. Maybe cows liked to face into the wind and the aroma of a distant feedlot. Perhaps they were being pulled to the sun, a phenomenon biologists like to call "phototropism," which is Greek for "turn" and "light" and thought to be restricted to plants and patrons of tanning salons. Phototropic plants bend toward the sun because, as Charles Darwin and his son Francis first discovered in 1880, the growing tips of the leaves and grass seedlings turn and grow toward a light source. Well, that's fine for plants, but cows don't come with growing tips.

None of these universal explanations would do, because the cows were not exhibiting universal directional behavior. Different groups of cows in different fields faced in different directions, some toward the sun, some toward Parliament and some toward nowhere in particular. The answer lay in the ancient artificial origin of cows—domestication—and one of its desired by-products: Cows are dumb. They follow one another, like sheep, who are also dumb, courtesy of domestication. A group of cows ruminating across a pasture tends to face in the same direction because they are all slowly following the lead cow, blindly grazing as they go. If the lead cow munches westward toward Topeka, so will all the others.

Cattle never played follow the leader quite so wimpily in the heyday of predomestication, before man transformed the wild, independent Ice Age aurochs (pronounced "oar-ox"; plural, aurochsen) into today's docile blimps of beef, the domestic humpless cow. During the Pleistocene (Ice Age) and Holocene (past 10,000 years), the species of wild aurochs, *Bos primigenius*, ranged over most of the Northern Hemisphere except North America. Cro-Magnon cave paintings in France and Spain depict ebony-black bull aurochsen standing six and a half feet at the withers with long forward-curving horns and white muzzles; females and calves were reddish-coated.

Domestic cattle *(Bos taurus)*, in contrast, have much smaller skeletons and less spunk. Man bred designer cattle from the aurochs to be easily herded and handled, namely, to be short in stature and smarts. Smaller may mean less meat, but a large, feral, uncontrollable cow is no meat at all. The breeding instructions are straightforward: Mate just the smaller, duller individuals in each generation. By the Iron Age, some domesticated cattle had been shrunk-bred to a height of three feet at the shoulder.[1]

Despite the onset of cowculture, the European aurochsen survived as a wild species during the Holocene alongside their domesticated brethren but dwindled in number at the hand of human hunters and the plow of agricultural expansion. They became extinct in Britain during the Bronze Age but hung on in central Europe until 1627 when the last known aurochs went down in a forest in Poland. The Indian variant of the wild aurochs *(Bos primigenius namadicus)* also succumbed to extinction after it was domesticated into the zebu, the Indian humped cattle *(Bos indicus)*.

Other wild species of Asiatic cattle have fared marginally better. The gaur *(Bos gaurus)*, the wild progenitor of the domesticated mithan *(Bos frontalis)* of Assam, Burma and Nepal, still browses the forests of India and southeast Asia. The same holds for the wild banteng *(Bos javanicus)* of Java and the wild yak *(Bos mutus)* of Tibet, Nepal and the Himalayas, although both have declined drastically in number and are now endangered species. They are the ancestors of their domesticated nemeses in the competition/extinction game, the domestic banteng (or Bali cattle) and the domestic yak *(Bos grunniens)*, respectively. The wild kouprey *(Bos sauveli)* of Cambodia, also a few bullhorns away from extinction, was never domesticated.

When man first decided to domesticate the aurochs, he couldn't just take the bull by the horns, drag him into camp and expect the rest of the herd to follow. Instead, man probably lured the animals to his campsites and eventually into his corrals by providing drinking water and especially salt, much as villagers do today with mithans in Assam. Initially, the captive aurochsen would not have been eaten—even early man knew better than to dip into the permanent endowment. Also, the aurochsen would not have been tame enough for milking. Rather, as in some current cultures, they were most likely a currency of barter in the Holocene commodities markets. Much later, during feudal times, this currency evolved into the Norman French word *catel* and its domesticated descendant *chattel* (or *cattle*), meaning movable wealth or property.

Domesticated aurochsen first surface at an archaeological site at Catal Huyuk in Turkey dated at 6400 B.C., where five-hundred years later there is evidence of a shrine with aurochs-related rites and icons. After 4500 B.C., different breeds of cattle—lyre-horned, polled (naturally hornless), longhorns and others—were commonplace in the Middle East and Mediterranean areas as draft, dairy and food animals, as sacred cows and as subjects of artistic-cum-erotic symbolism. Domesticated zebu (humped cattle) first appear on seals and carved stone vases as early as 2500 B.C. in Sumerian and Babylonian sites in Iraq and in the Indus Valley.

The cow stampede into the New World began in 1493 when Columbus, on his second voyage to America, fired the first volley in what was to become a transatlantic war of the worlds—Old World versus New, each sending its native animals and plants at the other. Columbus arrived at the Caribbean island of Española with a flotilla of "seventeen ships, 1200 men, and seeds and cuttings for the planting of wheat, chickpeas, melons, onions, radishes, salad greens, grape vines, sugar cane, and fruit stones for the founding of orchards."[2]

Also on board was "the first contingent of horses, dogs, pigs, cattle, chickens, sheep, and goats" to disembark in the New World.[2] The invading livestock throve on a continent rich in natural vegetation and poor in predators. Cattle multiplied so rapidly that thousands of head bulled their way across the continent ahead of the European settlers and reverted to the wild state. On northern Española, marooned sailors lived off the wild cattle; they smoked the beef on a wooden grate called a *boucan*, a word that evolved into "buccaneer" when the sailors turned to piracy in the 1600s.

American Indians, who were much better farmers than animal domesticators, welcomed the cattle and the new, inexhaustible supply of beef, hides and tallow. Until they came face to face with a made-in-Europe cow, the Indians' only other sight of a bovine was the American bison *(Bison)*, which later was hunted almost to extinction but never domesticated.

Artificial breeding, like evolution, leaves telltale pedigree trails of genes and physical features in descendant breeds and species: Some breeds of cattle retain the horn shape, or coloration, or stature of the ancestral wild aurochs. Armed with this principle, two German biologists, Lutz and Heinz Heck, set out to annul the eight-thousand-year course of cow husbandry: They aimed to re-create the extinct aurochs through artificial breeding, this time selecting for physical features that matched the cave paintings of *Bos primigenius* and the drawings of the last of the species that died in Poland in 1627.[3]

In 1921, Heinz Heck, director of the Munich Zoo, began crossing Hungarian and Podolian steppe cattle, Scottish Highland cattle, gray and brown Alpine breeds, piebald Friesians and Corsican cattle. By 1951 he had bred a herd of forty living "reconstituted aurochsen" that were the ripsnorting image of the extinct aurochs. At the same time,

Lutz Heck, working in the Berlin Zoological Gardens, also backtracked the genetic road to *primigenius* by crossing Spanish fighting cattle, Camargue cattle, Corsican and English park cattle. Most of his herd of reconstituted aurochsen was lost during World War II, save for a handful in European zoos and, ironically, a few individuals that had been released in a forest in eastern Poland.

The Polish version of the imitation aurochs, like the Munich one, is immense—a physical twin of its extinct wild ancestor. Remarkably, the beasts have also "reinherited" the wild species spirit: They are fierce, frisky and temperamental. They're also independent: In terms of the Cow Law, the only time the Polish reconstituted aurochsen all face in one direction is when they point their hindquarters toward Moscow.

My friend Zane Fross in Lost Cabin, Wyoming, works for the Spratt Ranch, a large spread along Badwater Creek and beyond. When I told him I had discovered the Cow Law, he informed me my discovery wasn't worth a bull's breakfast, which in this part of the country means a straw hat. Every rancher and farmer who raises cattle, from Wyoming to Zimbabwe, knows that a group of cows in an open field tends to face in one direction. Apparently, it's only a revelation to the gentleman farmer, who never raises anything but his hat, and to the paleontologist on I-80.

Undeterred, I told Zane there was a bushel more of I-80 cow tips where the Cow Law came from, which, if implemented, could revolutionize modern ranching. For example, why don't ranchers affix radio-tags to their cattle and keep track of them electronically, much as biologists do with bear, elk, pandas and other wildlife. Then the rancher can take his boots off, sit down in the den in front of his electronic cattleboard, and watch the blips graze across the screen, bulls in blue, cows in red, steers in pink.

When one wanders afield, a tiny shock will bring the blip back to the herd.

It's certainly better than branding, which was fine in the good old days of the Chisholm Trail and horses, when anyone could ride up to a stray cow, eyeball the brand, and know what the hieroglyph meant or which ranch it belonged to. Nowadays, though, the cowpoke is in the pickup truck and the cows are far afield, behind miles of interstates, gravel roads, and long-running fences. You need a telescope to make out the brand, and a code book to decipher it. And once it's decoded, who do I call about the stray steer on the road in California brandishing "IM-OK-UR-OK" on the rump?

No. Instead of branding, why not paint the telephone number of the ranch on both sides of each cow in huge white or Day-Glo numerals? Then, when a cow strays across a cattle guard or fence onto the gravel road or interstate, you know who to call. Large ranches with many head of cattle could use an 800 number.

For that matter, why don't ranches that border the highway lease their cattle as highway billboards? These "billcows" would generate revenue and abide by Lady Bird Johnson's highway beautification program: They're organic and don't block the view. The advertisement could be painted directly on the cow or slung over the sides. Dairy billcows would naturally advertise Dairy Queen, Frusen Glädjé and the like. Beef billcows would peddle burger joints and other calfeterias. Steer billcows would promote tofu, Greek mythology and baroque choral works for castrati. Advertisers could keep the moving mooing billcows near the highway (and passing motorists) with strategically placed bales of hay. Of course, unequal commercial exposure would not be a problem—all of the billcows and advertisements would face in the same direction.

# Chapter 16

*American bison, the organic food to a Plains Indian in the 1800s. (From John D. Godman,* American Natural History, *Carey, Lea & Carey, Philadelphia, 1828)*

# Oat Cuisine

Organic food, like chicken pox, is something you have to endure once in your lifetime. As a child, before we were beset with "natural food" emporia, I had to endure a daily dose of cod-liver oil, which was then hailed as the elixir of health and appropriately tasted like old sewer water.

Apparently not much has changed, at least not according to my dining experience. Foul taste, it seems, is still at the root of good health. At "organic" food stores, for example, granola, seaweed and tofu come packaged with promises of Mother Nature's goodness and the flavor of hay bales. At "natural" food restaurants, which often do double duty as vegetarian eateries, breakfast brings "organic" eggs—for "organic" read "fertilized"—and a choice of embryos over easy or scrambled. Lunch usually combines leek soup with an asparagus or parsnip sandwich—a menu stolen from some Russian gulag. Dinner features curdled soybean masquerading as turkey, steak, pastrami, cheese and milk shakes. One might as well chew on a sauteéd tire.

Meals come with bread, which for eight-thousand years has been an innocent mixture of flour and water. Now, for some reason, it's implanted with specks of zucchini, carrots, spinach or some other bits of rabbit forage and touted as manna from heaven. The star of oat cuisine is, of course, alfalfa sprouts, ordinarily the fodder of cattle, sheep, deer and other ruminants.

One of nature's meals is enough. I'll take my eggs inorganic, my bread plain, and my protein on the hoof. This isn't so much a red-neck sentiment as a red-meat one. According to extremists smitten with organic food, the diet of modern man is inferior to the natural one of prehistoric man. Even if this were true, I'd rather be inferior than prehistoric. In any event, the elysian scene of Neanderthals or Cro-Magnons merrily gathering nuts and berries is shattered by the archaeological evidence of prehistoric feasts on mammoths and reindeer.

Closer to home, eyewitness accounts of the American West of the 1880s record that whenever possible American Plains Indians chose buffalo banquets over grain. After the buffalo kill, "feasting is then commenced, and kept up day and night until meat has become scarce, when another hunt follows . . ." and "one ordinary or large family will sometimes consume a buffalo a day." One witness at an Indian meal wrote: "Everything ends by eating [buffalo], as it is said, with the stomach unbuttoned." Plains Indians would apparently gorge themselves with buffalo meat when it was available and go hungry when it was not.

Even at Indian trading posts, where the food supply was more regular, grains and greens were snubbed in favor of meat. During 1805 at Alexandria Post "upwards of 70 people consumed, at least, 450 pounds of buffalo meat every day—an average of 6.43 pounds per person." At Fort Qu'Appelle in 1867 to 1868, "the daily allowance for each child was one-quarter and for a woman one-half that for a man, which was 12 pounds fresh buffalo meat, or 6 pounds dried buffalo meat, or 3 pounds pemmican, or 6 rabbits, or 6 prairie chickens, or 3 large white fish, or 3 large or 6 small ducks, besides potatoes and some milk for the children, and occasionally dried berries, with a weekly allowance of tallow or fat."[1]

Whew! So much for ancient organic cuisines. In any case, most "natural" pretensions are merely semantic hogwash. A herd of Black Angus is just as "organic" as a forest of seaweed or a field of alfalfa.

Semantics aside, more serious is the naïve presumption of nature's benevolence, that its produce of wild greens, grains and fibers grows untainted in some Garden of Eden. Biochemistry tells a more realistic story. Most plants in nature are full of natural poisons: alkaloids, terpenes, phenolics, glycosides. Plants have evolved these natural pesticides as a defense against being attacked or eaten by a host of bacteria, fungi, insects, lizards and mammals. When it comes to pesticides, plants play hardball. At the very least, the bitter taste of the chemicals repels potential herbivores. At worst, the toxic pesticides in plants can produce intestinal, kidney and liver damage, anemia, reduced fertility, cardiac instability, circulatory failure, genetic mutations, cancerous lesions and death.

Herbivores, in turn, have evolved their own array of sensory, digestive and biochemical solutions to deal with the plant poisons. Some can smell the difference between a good cud and a potential case of food poisoning. Others detoxify or neutralize the plant pesticides through the action of digestive enzymes and intestinal bacteria. Some insects, such as monarch butterflies, not only tolerate the poisons, they can store them in high concentrations and use them to repel bird predators.

Over millions of years the "eaten" and the "eaters" have been engaged in a slow, deadly, co-evolutionary game of one-upmanship, a kind of organic arms race. Each new adaptation in plants, such as the poisonous chemicals, is matched by counteradaptations among successful herbivores. The process is natural selection: Only those plant eaters with the right counteradaptations will be able to eat, survive, reproduce and pass on to the next generation the genes that code for the particular counteradaptations. This, in turn, raises the ante for plants—only those with even more powerful anti-herbivore defenses will survive to reproduce—and brings a new cycle in the evolutionary treadmill of the eater versus the eaten. The same treadmill runs the evolution of all organisms competing for limited resources, and shapes those parts of their anatomy—the eating, moving, digesting parts—involved in the competition. One biologist called the treadmill the Rat Race; a second, borrowing from Lewis Carroll, dubbed it the Red Queen.[2]

Man, as one of the eaters, is also a target of the toxic chemicals synthesized by a garden variety of plants.[3] Celery, parsnips, parsley and figs produce natural mutagens and carcinogens called furocoumarins. Their potency increases with bruising or disease. Bruised celery, for example, causes skin rashes on the arms of celery pickers and handlers. Chemical agents in chocolate and cocoa (theobromine), potatoes (glycoalkaloids) and rhubarb (quinones) have proved toxic to laboratory animals, as have the natural pesticides in the fava bean, horseradish, oil of mustard, unrefined cottonseed oil and oil of sassafras—often an ingredient in "natural" sarsaparilla root beer. Extract of black pepper, in large doses, may cause tumors in laboratory mice; so do chemicals produced by the common commercial mushroom and its kin, the false morel. Okra and kapok contain toxic fatty acids. Herbs and herb teas synthesize toxic alkaloids. Beets, celery, lettuce, spinach, radishes and rhubarb produce natural nitrates, chemicals linked to some cancers. And alfalfa sprouts—yes, these too—may induce severe lesions when fed to laboratory monkeys.

On the surface this is grim news to patrons of organic cafés or, for that matter, anyone with a grocery list. It becomes black humor when we learn that humans consume at least 10,000 times the amount of natural pesticides as of man-made ones.

But beneath the prospect of a bleak palate lurks a safe meal. First, the results of many of the laboratory experiments are hotly contested among biochemists. Second, humans, fortunately, are not laboratory rodents, except when they dress that way for television game shows. Basal metabolic rate is much lower in humans than in rodents and, by implication, so is the potential toxic effect of plant poisons. And, like other herbivores, we too have evolved natural defenses against the pesticides in our food plants.

A more basic question is whether experimental animals, such as mice, rats and hamsters, have been artificially bred to such a degree that their responses to biochemical tests in the laboratory no longer mimic nature. It is possible that years of controlled breeding may have stripped laboratory animals of the defenses that natural selection imparted to man and other organisms in the evolutionary arms race. If so, that chemical standoff between the eater and the eaten—the result of millions of years of co-evolution—may no longer be coded for in the artificially selected genes of some experimental animals. A biochemist I spoke to considered this suggestion highly unlikely, but admitted that no one really knows.

One questionable element in the design of these laboratory experiments is known: the huge dosages administered to the laboratory animals. The act of overfeeding may itself be destructive.

Whatever the experimental results, two conclusions are unavoidable. First, organic menus are as suspect as square meals. Second, whether the cuisine is *haute* or oat, evolution has not played favorites with the ingredients. If you were to follow all of the experimental results in choosing a noncarcinogenic diet, you'd be eating oatmeal topped with bran flakes. Instead, I prefer the culinary wisdom of baseball players. When faced with the prospect of playing on Astroturf, first baseman and slugger Dick Allen remarked with typical rye humor: "If a horse won't eat it, I won't play on it."

# Chapter 17

*The hyrax is an example of an animal that could have been designed by a camel. It looks like a cross between a rhinoceros and a rodent and has been considered an evolutionary relative of elephants or sea cows or horses or rhinos or tapirs. (From William Jardine,* The Naturalist's Library. Mammalia, *Vol. V:* Pachyderms, *W. H. Lizars, Edinburgh, 1836)*

# Designed by a Camel

It is common knowledge that our common bullfrog can leap nine times its length. The question is, which bullfrog does the leaping? Zoologically, the bullfrog is classified as a large frog in the genus *Rana.* But in colloquial taxonomy, bullfrogs also go by belly-deep, bizmaroon, blood-lick, bloodynoun, brother-rounds, bull-paddock, bull-tucker, bullyrum, clunker, cowfrog, drummer, French frog, goodadoon, grandaddy, green frog, gurrump, honker, Irish nightingale, Johnny bull, jugarum, lunker, more-rummer, squawker, wart frog, yellowthroat.

Taxonomy, the art and science of classification, is a strictly human preoccupation. In biology, taxonomy is preoccupied with the orderly classification of nature. The exercise is hardly objective. In the workaday world, for example, nature is usually classified as follows: man, and the lower animals. Man, of course, is doing the classifying. Taxonomic objectivity increases in the scientific world, if only because science has failed to prove that man descended from a lower animal. Objectivity, however, brings complexity. Man is the genus *Homo*, *Homo* is a member of the family Hominidae, hominids are in the order Primates, primates are in the class Mammalia, mammals are in the subphylum Vertebrata and so on, a hierarchy based on the physical likeness, genealogical kinship and evolutionary history of animals.

Man, *Homo*, is a primate with a 2-million-year evolutionary history. Going by the cultures and classificatory habits of modern hunter-gatherer tribes, it's safe to conclude that members of the different species of *Homo* were also preoccupied with taxonomy during that 2-million-year history. *Homo habilis*, *Homo erectus* and *Homo sapiens* dealt with their natural world by pigeonholing animals and plants into neat, orderly categories: bushes with trees, butterflies with moths, frogs with toads, shrews with moles and man with apes.

Well, it may sound like a peaceful if dreadfully dull process, but it isn't. Nature isn't always orderly, the pigeonholes aren't neat and taxonomy became a high-spirited science. Debates over which species ought to be grouped in which category have been known to escalate from quibbling to wrangling to downright ridicule. I once ridiculed a colleague in print for what I considered his disorderly classification of extinct primates. He returned the favor, saying my classification of the same animals had set primate paleontology back twenty years. We've shaken hands since, as good primates with a 2-million-year evolutionary history should.

Squabbles over interpreting and classifying fossil animals are to be expected. After all, paleontology doesn't deal a full anatomical deck: When we find the remains of the extinct beasts, most of their "ologies" are gone—their embryology, physiology, soft-tissue morphology and so on. All that is left to recover and study are the petrified "ologies," namely the animal's osteology (bones) and odontology (teeth). Given such scattered, incomplete clues, disagreements over the evolutionary history and classification of fossil animals are understandable.

You'd think then that all living animals would be easy to classify in an orderly fashion. Think again. True, all of their "ologies" are present, with all the anatomical clues of evolutionary pedigree waiting to be deciphered. But what often intervenes is evolution itself.

For example, one of its processes, called convergence, produces alluring but false clues of evolutionary closeness. Convergence is the independent acquisition of similar physical traits in unrelated organisms. It produced bipedalism in birds and man: Both walk upright and both stand on two legs. Yet bipedalism came to birds from a primitive dinosaur living 200 million years ago, while two-legged man evolved from a four-legged hominid only between 5 and 10 million years ago.

Bipedalism also evolved convergently in different groups of dinosaurs (tyrannosaurs and coelurosaurs, duck-billed dinosaurs, dome-headed dinosaurs), in some rodents (jumping mice), marsupials (kangaroos) and as a special running posture in some lizards. Convergence also brought flight to such evolutionary strangers as birds, bats, pterosaurs and insects. It bestowed the animal-as-tank theme unto such unrelated beasts as armadillos, turtles, armored dinosaurs (e.g., *Ankylosaurus*), extinct reptiles called placodonts and extinct fishes called placoderms. Convergence caused whales and sharks to have a similar body shape, although one is a mammal, the other a fish, and the two are separated by more than 400 million years of evolution.

Most convergent similarities are only superficial, literally skin deep, and are easily exposed as filial frauds. But some are so striking that they tempt, trap and lead even the experienced taxonomist down blind evolutionary alleys.

The reasons for evolutionary convergence are twofold. Reason one is the constraints of efficiency and design. Potential evolutionary pathways are curbed by the animal's lineage and structure: There are only so many ways to convert a barn into an opera house. Genetic, embryologic and engineering constraints dictate that the basic body shape and skeletal design of mammals, for example, can be modified only a certain number of ways and still achieve an efficient marriage of form and function. In nature, natural selection umpires this tension; it is continually whittling an immense production of diversity down to what works at that particular moment.

The second reason behind evolutionary convergence is the success of imitation. Convergence is bred by the capitalistic maxim: Imitation is the sincerest form of evolutionary flattery, which explains why all chains saws, computers and CEOs look alike, no matter the company or country of origin. In nature, anatomical imitation is also a by-product of natural selection whittling across evolutionary lines. Over millions of years it produced thin, elongated limbs in horses and deer running from the same predators; it selected for similar, high-crowned, crested cheek teeth in distantly related grazers such as zebras, gazelles and rodents competing for the same food resource. In other words, natural selection often leads close ecological competitors down parallel evolutionary paths when survival depends on the same economy of form, function and successful design.

Convergence isn't the only cause of classificatory conundrums. Other taxonomic enigmas among living species come from opposite ends of the evolutionary spectrum: conservatives and extremists. Conservative animals have changed so little over millions of years from their ancestral condition that there are virtually no evolutionary links to tie these relics (sometimes called "living fossils") to living contemporaries. Evolutionary extremists, on the other hand, have specialized and changed to such a degree that their bizarre anatomy bears little trace of whence it came in the geologic past.

An example of an evolutionary conservative is the living tree shrew, a squirrellike long-snouted mammal that inhabits tropical forests in southeast Asia. Native Malaysians call them tupai, a name they also use for squirrels and one that taxonomists borrowed for *Tupaia*, a genus of tree shrew. At the turn of this century, paleontologists and taxonomists narrowed down the candidates for tree-shrew relatives to primates or insectivores. Tree shrews were assigned either to the order Primates (along with hominids, apes, monkeys, lemurs, lorises and tarsiers) or the order Insectivora (along with hedgehogs, shrews, elephant shrews, moles, golden moles and tenrecs). Either choice was an improvement over Linnaeus's classification of 1758. He had united hedgehogs, moles and shrews in an order Bestiae together with pigs, armadillos and opossums. All of these animals have an elongated snout, a convergence that fooled the father of taxonomy.

The tree-shrew debate raged until the early 1970s, with *Tupaia* swinging on the evolutionary tree between primates and insectivores. Then paleontologists solved the problem by relegating the dizzy tree shrews to their own mammalian order, the order Scandentia. Great. But what did it solve? On the surface, the classification of primates, insectivores and tree shrews was now orderly, each group cubbyholed in its own unique order. But the tidy classification merely masked the old knotty evolutionary puzzle: Which group had closer genealogical ties to tree shrews. Assigning the tree shrews to their own order was an admission that we couldn't choose between primates and insectivores, and until we could, tree shrews were best kept from sullying the classification of either. In taxonomic vocabulary, classifying tree shrews by themselves kept Primates and Insectivora from being taxonomic wastebaskets.

The problem with tree shrews is twofold. First is their comparatively bland morphology. Nothing about them seems to stand out as a genealogical beacon or evolutionary signature. They resemble archaic, extinct mammals, especially the primitive ancestors of insectivores and primates, which in turn lived over 65 million years ago and can barely be distinguished by their fossil remains. Tree-shrew teeth are a throwback to those extinct insectivores, whereas their foot bones are more like primates and bats. The second problem is their terrible paleontological record. Only two fossil tree shrews have been found, and they are far from the proverbial missing links.

Opposite tree shrews, at the other end of a disorderly classification, sit the extremists, those animals so specialized in anatomy that they've outstripped their evolutionary roots. A good example are the hyraxes, also known as conies, dassies or damans. They are hare-sized animals that live on rocky terrain throughout the Middle East and Africa. Their habitat earned mention in the Old Testament: "The conies are a feeble folk, yet make their houses in the rock" (Proverbs 30:26).

Forget feeble folk. If camels were designed by committee, hyraxes were designed by camels. They resemble a cross between a rhinoceros and a rodent. Their front teeth are chisellike and evergrowing, as in rodents, but their cheek teeth could have come from a miniature rhino. The forelimbs have four toes (like many rodents) that end in hoof-shaped nails (like rhinos); the hind limbs have three toes (rhinos), one of which ends in a long claw (rodents), the other two in hooflike nails (rhinos). Like rhinos and rodents, they are herbivores, feeding on grasses, leaves and bushes, and are basically short: They have short feet, a short snout, short ears and short hair. Otherwise, they're odd among plant eaters in retaining canine teeth and an extra bone in the wrist, and in having a gland on their backs marked by a patch of light-colored hair.

Such a smorgasbord of physical traits earned a dyspeptic taxonomy. Hyraxes have been classified with elephants, sea cows, horses, rhinos, tapirs as well as some extinct oddball mammals called embrithopods and desmostylians. Recent opinion is divided between a horse-rhino-hyrax evolutionary connection and a sea cow-elephant-hyrax linkage. Both were deduced from peculiar but different resemblances in the anatomy of the skull and foot bones.

Besides tree shrews and hyraxes, there is a passel of taxonomic nightmares among living mammals: The tarsiers (order Primates) are evolutionary conservatives; the sea cows (order Sirenia), aardvarks (order Tubulidentata), pangolins (order Pholidota), elephant shrews (order Macroscelidea) and colugos (order Dermoptera, "flying lemurs") are all evolutionary extremists. Some of these groups are living refutation of a benevolent Creator: They have all the humpmarks of more camel design.

The so-called "flying lemurs" don't fly and aren't lemurs. Their colloquial name is "colugos," they live in tropical forests in southeast Asia and their closest evolutionary kin are probably primates or tree shrews or bats or insectivores. Colugos "fly" from tree limb to tree limb like gliding squirrels, courtesy of an evolutionary specialization called a patagium, a parachutelike fold of skin that stretches from the forefeet to the hind feet. They also have peculiar comblike front teeth and multicusped cheek teeth. Some 55-million-year-old fossil sites in the Arctic and western North America yield strange jaws and teeth that may belong to ancient colugos. Then again, maybe they don't. In the meantime, colugos, like tree shrews, have been dumped into their own order, the Dermoptera.

Finally, the taxonomic nightmare becomes double-barreled when the spiny anteaters and duck-billed platypus (order Monotremata) of Australia and the sloths, anteaters and armadillos (order Edentata) of the Americas waddle onto the evolutionary stage. These groups are archaic conservatives in some parts of their anatomy, and outlandish extremists in others. Monotremes still lay eggs and have primitive shoulder blades, collarbones and ribs—lingering ghosts from a reptilian past. But they've also lost their teeth and evolved a spiny skin, a tubular snout and a ducklike bill, radical specializations found in few other mammals. Sloths, anteaters and armadillos are a similar anatomical patchwork of archaic retentions and dazzling novelties: primitive ribs and skull bones, poor control of internal temperature, reduction and loss of teeth, and extra bony protuberances on some vertebrae.

If the state of mammalian taxonomy sounds bleak, it really isn't. Paleontologists are practiced at sieving the mosaic checkerboard of traits in mammals for the evolutionary innovations that imply common ancestry, close kinship and neighboring pigeonholes in a classification. There are forty or so orders of mammals (twenty-four living, sixteen extinct), depending on who is doing the classifying. Of the forty orders, the unifying traits and relationships of twenty-four, or 60 percent, are basically sieved and settled. There are only two things paleontologists can do about the remaining sixteen unsettled mammalian orders. One, study more of their living species and find more of their fossil ones. Or two, call time out to consult the camels.

# Chapter 18

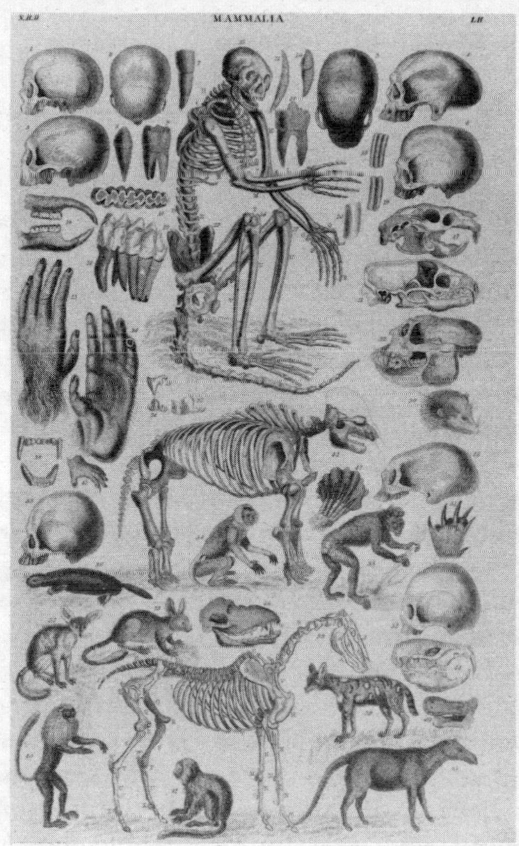


*Some of the players, living and extinct, in the mammalian hall of fame in 1840: man, monkey, mammoth, mastodon, marsupial, mouse, rat, lemur, lion, otter, orangutan, hyena, horse, duck-billed platypus. (From Oliver Goldsmith,* A History of the Earth and Animated Nature, *Blackie & Son, Glasgow, 1840)*

# Life from the Cheap Seats

All the seats in old Delormier Stadium in Montreal were cheap. Half a buck bought you nine innings of the Montreal Royals, and a preview of the future Flatbush bums. Jackie Robinson. Duke Snider. Junior Gilliam. The Royals were a farm club of the Brooklyn Dodgers.

One afternoon, when I should have been in school, Gilliam smacked consecutive home runs on two pitches, both a foot on the wrong side of the right-field foul pole. Both broke windows on the fifth floor of a battered warehouse behind the stadium. A few years later baseball went broke in Montreal. The stadium, the Royals, even the warehouse became extinct. The Dodgers migrated from Brooklyn and Ebbets Field to a new habitat in Chavez Ravine.

Baseball imitates life. At least it will start doing so in April, when baseball shuts down the Grapefruit League, migrates north and laces up for opening day. Come October, the end of the pennant races, the World Series and the beginning of cold weather, the ball clubs lead a bevy of creatures into hibernation. The teams snooze till March, when they emerge with the animals and begin spring training. The Pittsburgh Pirates often hibernate until the All-Star break.

Hibernation, an adaptation to cold, allows warm-blooded animals, such as mammals and birds, to avoid a deadly biological double-play: excessive heat loss and a scarce food supply during the frigid winter months. Hibernating animals also go into physiological slow motion. Body temperature plummets; so do heart rate, metabolism and respiration. In hibernating ground squirrels, for example, body temperature falls from a normal 98 degrees Fahrenheit to 35 degrees, and heart rate from about 350 beats per minute to only 2 or 4. They survive the winter in this dormant state, slowly tapping their own energy reserves. Bears hibernate, which comes as no surprise to Cubs fans. What is surprising is that females will give birth to cubs while still in hibernation during late January or early February. The newborn cubs, weighing about half a pound, then suckle and grow while mama sleeps.

Baseball, like life, also pays an evolutionary price: periodic extinction. Witness the Boston Braves, Brooklyn Bridegrooms, Philadelphia Athletics, New York Highlanders, Houston Colt 45s and Seattle Pilots, only a few clubs on baseball's roster of extinct species. The Washington Senators have the dubious distinction of having belly-upped twice, a feat unmatched by any other member of the animal or plant kingdom. The first species of Senators evolved into the Minnesota Twins; the second Senator team died and was reborn as the Texas Rangers. Perhaps future Washington baseball clubs will avoid the jinx of a congressional moniker.

Although "extinct," the players, teams and ball parks are preserved in memorabilia: baseball cards, old programs, faded pennants and other petrified remains of fossil franchises. Cooperstown's Hall of Fame—baseball's version of Dinosaur Hall at The Carngeie Museum—is crammed with these relics of ancient life on the diamond.

A vital difference, though, between biological extinctions and those that occur in baseball is finality. Evolutionary resurrections are apparently fair in baseball, but they are foul balls in nature. Millions of plant and animal species have come and gone, but none have been reborn: dinosaurs, dodos, mastodonts, mammoths, moas, pterosaurs, passenger pigeons, ammonites and sabertooth cats have vanished for good. In baseball, on the other hand, real extinction has claimed only a few ball clubs, like the Montreal Royals. Otherwise, most baseball extinctions have been local disappearing acts, what paleobiologists like to call "extirpations." Teams like the Dodgers and Giants merely pulled up home plate, moved to a new town and put down grass in a new stadium.

Some teams pulled this extirpation act twice. The Boston Braves became defunct in Beantown and migrated first to Milwaukee and later to Atlanta. The Athletics, after suffering local extinction in Philadelphia, gradually colonized the West by conquering Kansas City and then Oakland. They should have changed their name to the Oakland Mastodonts, in honor of the all-world migrators. Starting 25 million years ago, mastodonts spread from Africa into Europe and Asia, then across the Bering land bridge into North America and, finally, about 2 million years ago, invaded South America. Early man followed the mastodonts 24 million years later when *Homo erectus* meandered from Africa into Europe and Asia; about 15,000 years ago *Homo sapiens* crossed from Asia into North America and spread to Central and South America.

After suffering local extinction, some teams undergo a modicum of evolutionary change—what in evolutionary language is called "speciation." The teams migrate to a new locale, change their name and uniform, and evolve into a "new" species. For example, the Seattle Pilots fled the northwest and "speciated" into the Milwaukee Brewers. But this isn't real speciation. It's fake or pseudo-speciation—after all, the ball club is the same, only its name has been changed to protect the bankrupt. In New York and Houston, pseudo-speciation occurred in situ: The New York Highlanders became the Yankees, and the Colt 45s begat the Astros.

Biologically speaking, the "new" team is merely a renamed segment of one continuous, evolving lineage. Among organisms this is called sequential evolution or anagenetic change or phyletic gradualism, a process of steady, incremental anatomical change recorded in the fossil record of many animals, including early man. One major reconstruction of the pace and pattern of human evolution is the sequence *Homo habilis* to *Homo erectus* to *Homo sapiens*, each species grading into the next in such features as increasing brain size, larger body size and a higher-domed skull. Indeed, the overlap in physical attributes often blurs the border between adjacent "pseudo-species" in an evolutionary lineage. In these instances, their fossil bones confirm the pasty cliché "bones of contention," being as easily placed in one species as the other.

True speciation, on the other hand, is clear-cut. A species is a species because its members interbreed. But sometimes outlying or far-flung members are cut off geographically or ecologically from the main body of the species. If they can't meet, they can't mate. Given enough time, these isolated groups, which biologists call "peripheral populations," will evolve in different directions to a point where they and their former mates are breeds apart—sexually incompatible and physically distinct.

This budding off of new species is called allopatric speciation in natural history and expansion in baseball. A good example is the origin of the Montreal Expos and the Toronto Blue Jays from peripheral populations of other teams. Currently, Denver, Vancouver and Tampa-St. Pete are trumpeting their geographic isolation and unique genes in the hope of allopatrically speciating a new baseball franchise.

Of course, some teams just don't evolve much. They don't migrate, don't speciate, don't even pseudo-speciate. Instead, they show stasis over long periods of time: the Detroit Tigers, St. Louis Cardinals and Pittsburgh Pirates are hallmarks of stability. This stability has been outmatched, however, by most animals. Extreme examples are the so-called "living fossils": *Lingula*, a lamp shell, which has survived for more than 400 million years with little apparent restyling; and *Latimeria*, the living coelacanth, which is remarkably similar to its 100-million-year-old extinct relatives. A good many species in the fossil record, from snails to opossumlike marsupials, appear to have survived virtually unchanged for 5 to 10 million years.

Every armchair manager knows that survival in baseball is a matter of consistency and diversity. The teams that consistently produce more runs than they allow survive to make the playoffs, the World Series and a healthy profit. Diversity helps. Smoke-throwing relievers complement the finesse pitcher. Clubs with a mix of power hitters, base stealers, righties and southpaws will have a natural advantage in the competition on the diamond.

Darwin would have made a great baseball manager; he observed the same principles of consistency and diversity in the natural world. Animal species survive if more of their individuals consistently reach maturity (reproductive age), mate and produce offspring than do competing species. Here too diversity is critical. Thanks to sex—and its reshuffling of male and female parental genes—a species consists of a diverse lot of individuals, differing from one another in genetic and physical makeup. Some of these differences (or traits) are favorable: They confer a natural advantage in the competition for food, space and mates in specific environments. Individuals with more favorable traits tend to be "selected for" in the game of natural selection—they have higher odds of reaching sexual maturity, mating and passing on their favorable genes to the next generation, where the same game is repeated. Natural selection promotes a hierarchy of success: from favorable genes, to the physical traits the genes code for, to the individuals carrying those traits and, over time, to the species those individuals compose. Whether a species as a whole survives (is selected for) or disappears (is selected against) depends on the grand tally, the line score of the life and death of its individuals over time. Winning species endure for the moment; losing ones peter out. But should the environment change drastically, what was a "favorable" trait may go out the Darwinian window, and survival suddenly becomes a whole new ball game.

The ups and downs of environmental change aren't news in the dugouts. Teams that reigned on natural turf became Keystone Kops on the faster artificial carpet. Players became extinct on Astroturf after suffering the kind of injuries that would never have occurred on grass. Baseball has a full count of environmental variables that affect each game and survival: umpires, stadium dimensions, fans, corked bats, Vaseline balls and so on. Teams alter the environment to their advantage, shifting the home-run fences for the sluggers, shaving the grass for the bunters and doctoring the ninety feet between first and second for the base stealers. With the shift from grass to Astroturf, modified anatomy evolved: Cleated shoes gave way to ones with ribbed soles.

Ditto for hoofed mammals, which have undergone a

similar evolutionary change over the past 40 million years. With the slow shift over the earth from tropical forests to grasslands, horses, for example, became "reshoed": Three of their original four toes on each foot gradually disappeared; the remaining toe, the third, enlarged to form a single broad hoof, an adaptation for speed. Rhinos and tapirs went from four toes to three; pigs, deer, antelope, cattle, sheep, goats and most other artiodactyls (even-toed ungulates) went from four toes to two: the "cloven hoof."

Like so much of geology, baseball is propelled by continental drift: Basically, the American League and National League are separate, floating continents that collide every October. Only then, during the World Series, do the leagues mingle umpires, fans, ball parks, strike zones and designated hitters. Otherwise the two leagues are distinct land masses, their teams playing and evolving in isolation.

Animals and continents have been at this for over 200 million years, ever since the breakup of the supercontinent Pangaea into drifting island continents. For example, after North and South America became separated by the Isthmus of Panama about 80 million years ago, land-bound animals on each continent were marooned and evolved in isolation. North America became the main dominion of the nonpouched mammals (placentals), a floating National League franchised with carnivores, rodents, rabbits, shrews, primitive primates and a horde of hoofed mammals, such as horses, tapirs and rhinos (perissodactyls or the odd-toed ungulates), and pigs, camels and deer (artiodactyls).

In contrast, South America was left mostly with a passel of marsupials (pouched mammals), and only one or two placental stragglers that had arrived in South America from the north prior to the split. The marsupials flourished in the South American League, evolving ecological counterparts to many of their North American placental cousins: marsupial "dogs," marsupial "cats," marsupial "shrews," marsupial "rodents," and so on—essentially the same players, but on a different continent and dressed in different anatomy. The few placental stragglers in South America diversified into today's armadillos, sloths, anteaters and a bizarre array of extinct hoofed herbivores with names that no baseball team has yet adopted: notoungulates, litopterns, astrapotheres, pyrotheres. The Pittsburgh Pyrotheres would be a fitting name for a second franchise.

For 77 million years, the bestiaries of North and South America played and evolved in isolation, although there were some minor episodes of league jumping. About 35 million years ago, some primates and rodents, in an imitation of free agents, managed to island hop or raft to South America from North America or Africa. Also during this period, North America traded mammals with Europe and Asia across northern Atlantic and Pacific land bridges.

The north-south isolation ended abruptly about 3 million years ago when, driven by the geologic forces of continental drift, the two Americas collided and reestablished a land link. The Great American Interchange, geology's World Series, was on. Swarms of North American placental mammals streamed into South America: rabbits, shrews, mice, squirrels, dogs, cats, raccoons, weasels, horses, tapirs, peccaries, deer, llamas and mastodonts. At the same time a few South American mammals went north against the flow and spread into Central and North America: opossums, porcupines, capybaras (gigantic rodents), armadillos and their now extinct giant cousins, the glyptodonts.

When the dust settled, it was obvious the North American mammals had won. The native South American beasts—both the marsupials and placentals—came out on the short end of the Great American Interchange. By 3 million years ago, both had already dwindled in number and diversity. With the invasion of the northern placental competitors, they succumbed to extinction.

Baseball endures because it is timeless—it is measured in outs, not hours. But baseball in the Big O, the new Olympic Stadium in Montreal, is unendurable. For one thing, the seats are no longer cheap. Nine-fifty will buy you a spot in the upper deck, closer to the asteroid zone than the Astroturf. You get a blimp's glimpse of Montreal's new franchise, the Expos, in action. And *le baseball* (in French). *Le strikeout* won't do. Rather, it's a *retrait sur des prises* ("a retreat on strikes") and other mutations that characterize baseball's bilingual speciation event. No wonder Les Expos, although burgeoning with talent, are also les losers. After all, can you blame Gary Carter or Andre Dawson for giggling at the plate instead of concentrating on the pitch after hearing himself introduced as *le frappeur*? For Expos fans, this is a sorry evolutionary turn of events.

## Epilogue

Baseball continues on its evolutionary way in Montreal. Gary Carter and Andre Dawson no longer play for the Expos, having been dealt to the Mets and Cubs, respectively. Olympic Stadium, the "Big O," is also called Taillibert's Mistake, in honor of Roger Taillibert, the French architect who designed the stadium for the 1976 Olympic Games. Taillibert envisioned not a stadium but a "statement," the architectural epitome of Montreal's culture and soul, something akin to Paris's Eiffel Tower, or the Cathedral of Notre Dame. Montreal's statement ended up being a huge concrete flying saucer that Richie Hebner once described as the world's largest toilet bowl. Baseball people concede that Olympic Stadium is a stadium, but they don't know for what. Only the Montreal taxpayer knows—for debts. The Big O was rightfully dubbed the Big Owe, as it won't be paid off until 1994. It cost almost $1 billion to build, which is more than the seven other domed stadiums in North America combined. And the retractable dome, which was promised for 1976, was finally delivered in April 1987. It leaked during the first rainstorm, and as yet it can't retract. Otherwise, it works fine.

# Chapter 19

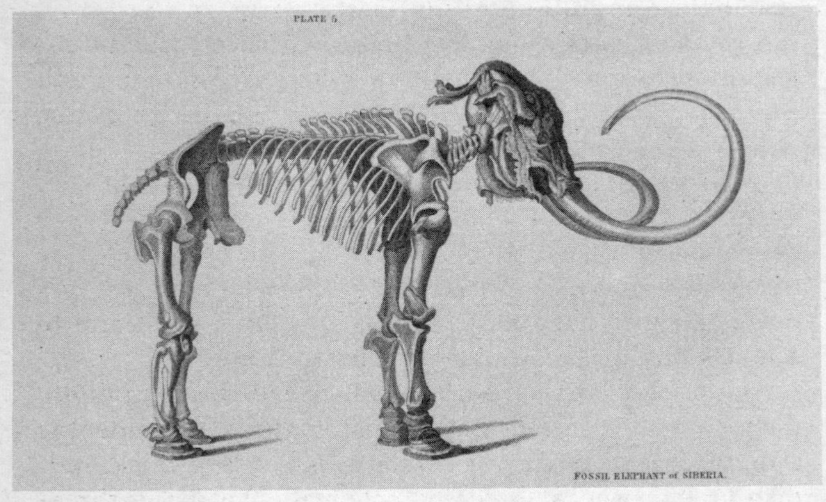


*This mammoth skeleton was discovered in 1799 on the shores of the Lena River in Siberia. It was excavated in 1806, hauled to St. Petersburg, and displayed in the Museum of the Imperial Academy of Sciences. (From William Jardine,* The Naturalist's Library. Mammalia, *Vol V:* Pachyderms, *W. H. Lizars, Edinburgh, 1836)*

# Mammoth Tales

In 1901, an expedition from the Russian Academy of Sciences arrived in the Siberian Arctic to excavate the carcass of an extinct mammoth. It had been found the previous year by a Lamut, one of the local natives, in the frozen soil of the tundra along the Berescova River. Lamut superstitions warned that tampering with a frozen mammoth would produce terrible afflictions, and this had kept the carcass from being exposed and dismembered. Expedition members were told of a Siberian who had died along with his entire family after he had examined a mammoth. But the taboo had been overridden by economics: The huge white tusks and their prized lode of ivory had disappeared down the merchant route of the lucrative ivory trade to Yakutsk and on to China or Western Europe.

As the expedition members unearthed the mammoth, the gigantic body of the beast began to thaw. The hide was a dense mat of long, coarse, black hair with a thick, reddish-brown, woolly undercoat. Beneath that lay a three-inch layer of white fat and, under the fat, pink muscle. The stomach held remnants of the mammoth's last meal: bits of twigs, bark, grass and leaves. Parts of the body had been gnawed by dogs, foxes, bears, wolves and wolverines.

What remained of the carcass and skeleton was packed in snow and ice and hauled back to the Academy of Sciences in St. Petersburg (now Leningrad) but, as rumor has it, not before the expedition members yielded to a singular paleontological temptation: to taste the flesh of an extinct animal. They trimmed off a few steaks, grilled them on an open fire and, like *Homo sapiens* hunters in Siberia tens of thousands of years ago, feasted on roast mammoth.

As the story goes, the meal made them violently ill, lending backhanded credence to Lamut superstitions. More realistically, perhaps cryogenics and exotic cuisine aren't everything they're cracked up to be. A second version of the story maintains that the mammoth was one of the delicacies served at that year's annual banquet of the Academy of Sciences, with similar dyspeptic results. Both tales, apparently, are mammoth fables, according to a young member of subsequent expeditions to Siberia, I. P. Tolmachoff, who later became curator of invertebrate fossils at The Carnegie Museum of Natural History and wrote a history of the discovery of frozen carcasses of mammoths and rhinoceroses in the Siberian Arctic.[1] He describes how the mammoth carcass was collected, stuffed and mounted in the Zoological Museum of the Academy, but never eaten, either in Siberia or in St. Petersburg.

Following the discovery and excavation of the Berescova mammoth, a string of tragedies struck, which only reinforced Lamut superstitions. The leader of the academy's expedition, a zoologist named O. Herz, died suddenly two years after the mammoth was unearthed. The Cossack who had reported the carcass to the local chief of police apparently went insane and drank himself to death. The chief of police, who had visited the locality and informed the Academy of Sciences of the mammoth find, died two days before he was to be rewarded by the tsar with a medal of honor.

## Mammoth Tales

A scattershot of fact often lurks behind animal legends and myths. The Lamut superstitions, for example, probably arose after some of the natives became ill or died from eating the putrefied flesh of a decaying mammoth. Other mammoth tales, however, smell more of figment than fact, especially the religious fables that abounded in Europe in the Middle Ages. In Greece and Rome, mammoth bones were revered as the remains of pagan heroes or Christian saints. In a church in Valencia, a mammoth tooth became the venerated molar of Saint Christopher, and in 1789 a mammoth thigh bone, honored as the arm bone of a saint, was carried aloft through the streets in a mammoth prayer for rain. In Scotland, the bones of mammoths were hailed as those of human giants, such as the fourteen-foot "Littel Johne" of Robin Hood fame. One tale at least had fossil proof on its side: Mammoth bones in the Pyrenees and Alps were considered the remains of some of the elephants that Hannibal and his army had brought from Africa in 218 B.C. for their long march on Rome.

The genuine mammoth tale begins in Africa, the evolutionary birthplace of elephants, mammoths, mastodonts and other extinct elephantine beasts—a group of mammals known as the Proboscidea, named for their most obvious appendage, the proboscis, or trunk. Colloquially, they're known as pachyderms (literally, "creatures with thick skin") or as the animals with a vacuum cleaner up front and rug beater in back. Basically, mammoths were elephants with a lot more hair. They are close evolutionary cousins of the living Indian elephant (*Elephas maximus*) and more distant kin of the African one (*Loxodonta africana*), the only other extant species of elephant.

Proboscidean antiquity reaches back at least 45 million years to a fossil beast from Egypt called *Moeritherium*. Although small in comparison to its descendants, *Moeritherium* already had all of the teeth, nose and ear marks of later elephants, mammoths and mastodonts: an enlarged set of upper incisor teeth (or tusks), a short trunk and stout limbs. *Moeritherium* weighed about 440 pounds; over the next 40 million years, some of its descendants would reach a weight of 9 tons and a shoulder height of fourteen feet.

For 20 million years after *Moeritherium*, proboscidean evolution remained confined to Africa and produced a dizzying array of trunked, tusked behemoths on that continent. One major group was the elephants and mammoths. A second offshoot was the true mastodonts and the so-called pig-toothed mastodonts, or gomphotheres. All attained a gigantic bulk, massive, pillarlike limbs to support that weight, huge, sinuous tusks, and large cheek teeth. With longer limbs evolved a longer, flexible trunk, which was their only access to the ground when foraging for food and water. Elephants stand about ten feet tall; their enormous skulls, armed with two 190-pound tusks, weigh in at more than half a ton. Without a trunk, elephants would have the impossible task of lowering and raising this massive head for feeding and drinking.

After their origin, mastodonts/gomphotheres on the one hand, and elephants/mammoths on the other went their separate evolutionary ways. Mastodonts retained a long, low skull, cheek teeth with rows of simple, conical cusps and a second set of enlarged tusks in the lower jaw. But in elephants and mammoths the cheek teeth developed large cross plates of enamel for chewing hard grasses and the lower jaw never bore tusks. Their skull was also revamped: domed in the back, high in the brow region and flat-faced, all shaped to hold huge swaths of muscles for supporting the head, trunk and tusks, as well as for operating the giant grinding jaws and teeth. This wasn't evolution's version of a beauty face-lift. Living elephants may be "cute" by consensus, but a bare, undressed elephant skull is disfigured, almost evil, a sinister mangle of bones and empty sockets. Here, I suppose, beauty is in the eye of the behemoth.

About 25 million years ago, mammoth, mastodont and elephant evolution expanded out of Africa when geologic activity built land bridges with Eurasia across the Middle East and the western Mediterranean. Groups of African pachyderms spread to Europe and Asia, across the Bering area to North America and eventually to South America. A plot of the successive wanderings of elephants, mammoths, mastodonts and gomphotheres across these continents during the last 20 million years reads like a Pan Am route map.

The first to arrive in North America were the mastodonts and gomphotheres, about 15 million years ago. Mastodonts were conservative and stayed put. After 13 million years, by the beginning of the Ice Age, only one species, the American mastodont, roamed the continent. It stood ten feet at the shoulder and had a coat of shaggy brown hair. Stomach remains from preserved carcasses indicate it browsed on spruce cones, twigs, leaves, coarse grasses, swamp plants and mosses.

Gomphotheres, on the other hand, were expansionists. They invaded Central and South America about 3 million years ago, when continental drift built the first land bridge across the Isthmus of Panama. These pig-toothers flourished in South America, diversifying into a herd of beasts with outlandish anatomy and outrageous names: shovel-toothed amebelodonts, beak-toothed rhynchotheres and short-jawed anancines. One of the last American gomphotheres was named *Cuvieronius*, after the French natural historian Baron Georges Cuvier, who cut his paleontological teeth on fossil proboscideans and coined the term mastodont.

Mammoths were latecomers to North America, but they made up for lost time by migrating twice across the Bering land bridge from Asia. The first invasion came about 1.5 million years ago, when the southern mammoth joined the American mastodont in North America to become the dominant Ice Age herbivores. They ruled the spruce woodlands and forests of the north and west, the prairies of the Great Plains, and the lowlands and swamps of the coastal plains. During the Ice Age, the southern mammoth evolved into the larger, bulkier Columbian mammoth and, in turn, into the enormous Jefferson's mammoth, named for president and part-time paleontololgist Thomas Jefferson.

The second mammoth invasion from Europe and Asia, about 100,000 years ago, involved a more famous species, the cold- and tundra-adapted woolly mammoth. Arriving in North America, the herds of woolly mammoths met up with Jefferson's mammoth and the American mastodont; this proboscidean triumvirate stomped over the continent until their sudden extinction closed out the Ice Age about 10,000 years ago. The woolly mammoths that stayed behind in southern Europe became favorite portrait subjects of Ice Age tribes. Exquisite drawings of these beasts were left by Cro-Magnon people on the walls of caves in southern France and Spain.[2] But, suddenly, 10,000 years ago, coinciding with the North American extinctions, the woolly mammoth also vanished from Europe.

Mammoth tales have long tied these extinctions to the biblical deluge. One story claims that the mammoths and mastodonts were too large to gain admittance to Noah's ark. Another holds that they managed to squeeze on to the ark's passenger list, but perished when, upon disembarking, they sank into the flood-soaked earth.

Biblical fables aside, the sudden demise of mammoths

and mastodonts was only the tip of the glacial extinctions: Entire mammalian faunas of North America, South America and Europe were decimated between 12,000 and 10,000 years ago at the end of the Ice Age. The animal tombstone for North America reads: all mammoths, mastodonts, gomphotheres, horses, tapirs, camels, ground sloths, glyptodonts (giant armadillos), peccaries, cheetah and sabertooth cat as well as individual species of bears, giant rodents, deer, musk-oxen and moose. South America lost various glyptodonts, sloths, rodents, carnivores, mastodonts, horses, peccaries, camels, deer and strange native hoofed mammals called notoungulates and litopterns. Europe witnessed the disappearance of rhinoceroses (woolly and "naked" species), the woolly mammoth, giant deer, hippos, musk-oxen, hyenas, antelope and elephants. According to some reckonings, 75 percent of all large animals (over twenty pounds) and 90 percent of all large mammals (or seventy genera) vanished from these continents.

The plot of this Ice Age animal annihilation has come down to either a "whatdunit" or "whodunit." Paleobiologists in the "whatdunit" school blame the wholesale extinctions on rapid climatic and environmental change at the end of the Ice Age brought on by the retreat of glacial ice. According to this scenario, temperatures became more extreme, devastating existing plant habitats and communities to the point where they could no longer support a diversity of plant eaters. These disrupted vegetational patterns, in turn, upset the longstanding, delicate balance in the ecosystem of grazing and browsing animals, perhaps affecting their reproductive behavior. Such massive upheavals of habitats, say the climate-blamers, led to the wholesale extinction of most of North America's large herbivores—mammoths, mastodonts, horses, tapirs, camels, grounds sloths—as well as their predators.

Mammoth bull, says the "whodunit" school, led by Paul S. Martin, an archaeologist at the University of Arizona. Martin lays the blame for the dramatic Ice Age die-off at the bloodied hands of early *Homo sapiens*, whose arrival in North America from Asia about 11,500 years ago coincides with the sudden extinctions. The Americas were then a hunter's "Garden of Eden" gorged with 50 million to 100 million large mammals, the kind of game the human immigrants had grown proficient at hunting in Europe and Asia.

Martin's hypothesis is straightforward overkill: Moving at no more than about twenty miles a year, a rapidly spreading front of Ice Age human hunters slaughtered, butchered and exterminated the big animals of the Americas, beginning in northern Canada about 11,500 years ago and finishing the job in Patagonia 1,000 years later.[3]

Both the climatic and overkill theories have their problems. There is no question that Jefferson's mammoth and the woolly mammoth were hunted by Paleo-Indians in North America. The woolly mammoth was also constant prey for Ice Age tribes in Europe. But if Martin is right and the hunting reached overkill proportions, why are there so few kill sites in the archaeological record? Armed and ready, Martin answers with an ad hoc spray of rhetoric he calls the "blitzkrieg" addendum: Overkill was so swift and devastating that few kill sites are to be expected. But how could Ice Age hunter-gatherer tribes devour such tonnages of meat? Answer: Humans were probably as wasteful then as now. What about the mammoths, mastodonts, peccaries, camels, llamas, ground sloths and sabertooth cats that apparently survived the human blitzkrieg 10,000 years ago?—their remains at some sites date to between 9,000 and 5,000 years. Martin, with fairly good justification, claims those radiocarbon dates are suspect, as they are based on bone collagen, an unreliable tissue for dating. How does one account for the many species of birds and mammals that were not prey for Ice Age hunters and still became extinct? Martin says they were scavengers that, because of the overkill extinctions, were robbed of their steady supply of carcasses.

Different criticisms haunt the climatic theory of Ice Age extinctions, especially their selective impact: Why, along with the large mammal extinctions, wasn't there equivalent devastation among plants and cold-blooded animals (for example, reptiles, amphibians, fish and beetles)? Why did the last glacial retreat across the Northern Hemisphere ravage the American faunas but barely scar the European one? The answers to these and other questions await a more accurate timing of Ice Age events and a better fossil record of the animals and plants involved.

Both schools, overkill and climate, bemoan the fact that, at this point, hard evidence for the exact timing of the Ice Age extinctions exists for only seven of the seventy vanished species of large mammals.[4] An emerging consensus from ongoing research is that the either/or apposition of human overkill versus climatic change is too simplistic. The lethal weapons of extinction may have been both, a one-two punch of cultural decimation and environmental change acting in different proportions on different animals and different continents.

For example, mammoths were especially vulnerable to overkill. They were herd animals, easy to slaughter in large numbers. Also, their breeding rate was extremely slow; judging from elephants, mammoths probably didn't reach sexual maturity for twenty years or more. They could never outreproduce the kill rate. The American mastodont, on the other hand, would not have been susceptible to overkill. Individuals appear to have roamed singly rather than in herds, and lived almost exclusively in dense spruce forests, an environment unsuited for massing and killing prey. Instead, it was probably the mastodont's ecological dependence on spruce forests that marked it for extinction: Climatic change at the end of the Ice Age ravaged the spruce-forest habitat, reducing it to patchy stands of trees scattered across North America amid glacial lakes, ice, shrub tundra and deciduous forests. Had the mastodonts been able to hang on for another thousand years, the species would have witnessed a resurgence of its old habitat—a new burgeoning belt of spruce forests stretching across northern Canada—where they might have survived the Ice Age scourge. Had they dodged extinction, mastodonts today would be venerated (and trophy hunted) as the largest living land mammal on earth.

An intriguing union of the blitzkrieg and environmental theories is the "keystone herbivore hypothesis" of Ice Age extinction put forward by Norman Owen-Smith, a biologist at the University of the Witwatersrand in South Africa.[5] His explanation begins with the critical ecological role played by "megaherbivores" such as elephants and mammoths, those plant eaters weighing more than two thousand pounds. The grazing and browsing of modern megaherbivores have an immense impact on plant growth, nutrient recycling, and the composition and turnover of vegetational habitats. Just by felling trees and feeding, megaherbivores normally transform wooded environments into grasslands, or create a patchwork of open glades in a dense forest; in Africa, for example, elephants and rhinos spur the conversion of wooded savannas to more grassy savannas. These vegetational transformations initiate an ecological domino effect: They create a host of varied habitats and new plant resources for smaller herbivores, which in turn affects the food supply of their predators.

Owen-Smith maintains that extermination, probably by human overkill, of mammoths and other megaherbivores at the end of the Ice Age started a deadly ecological chain reaction. Without the massive, steady vegetational cropping, vegetation patterns in the Northern Hemisphere were drastically transformed: Open forests became closed, dense and dark; shrublands grew wooded; and savannas reverted to grasslands. With their once widespread habitats now sparse and restricted to isolated pockets, the medium-sized plant eaters became extinct, followed shortly by their carnivorous hunters.

Ten thousand years after the mammoth's extinction, its folkloric name entered natural history. From the Siberian *mammantu* sprang the scientific names *Mammut*, the genus of mastodont, and *Mammuthus*, the genus of mammoth. Their petrified tusks became treasured by art and commerce: Most of the 25,000 fossil woolly mammoths discovered in Siberia lost their tusks to a booming ivory trade. Hundreds of tons were exported from Russia starting in the fourth century B.C. Siberian ivory reached southern Europe and central Asia by the tenth century, and Mongolia by the thirteenth. Mammoth ivory came to compose such treasured objects as the throne of the Great Mogul Khan Kuyuk in Mongolia, fine jewelry in France and pistol handles in America.

Of course, it was Ice Age man who first made fossil ivory a currency of artifacts and art. From mammoth tusks they fashioned spears, staffs, needles and other cavehold tools, many of them decorated with engraved images of horses, mammoths, bison and ibex. From chunks of ivory they sculpted the "Venus" figurines, the supervoluptuous odes to Stone Age women. And, most recently, we learned that Ice Age hunters in Poland 23,000 years ago may have whittled a mammoth tusk into the world's oldest boomerang.[6] While excavating a cave in southern Poland, scientists from the Polish Academy of Sciences in Krakow recovered a curved, two-and-a-half-foot length of ivory in the telltale shape of a boomerang: It is rounded along one edge and polished flat on the other. Found along with the purported boomerang were teeth and bones of mammals, reptiles, birds and amphibians, stone and bone tools of an Ice Age technology and the thumb bone of an Ice Age human boomerang-maker.

Archaeologists are skeptical of the boomerang attribution. They caution that a curved, flattened piece of tusk does not a boomerang make. Museum cabinets housing archaeological collections are filled with ancient curved "killing sticks" that look superficially like boomerangs but fly like cement ducks. The academy scientists aren't about to test whether the Polish "boomerang" can fly right. Boomerang or not, this prehistoric artifact is too valuable to risk whirling into space only to have it plummet to the ground and smash into bits. Testing a replica won't do either: Burial and fossilization of the "boomerang" may have damaged its delicate contours and altered its subtle aerodynamic properties.

More credible than a prehistoric Polish boomerang is the claim by one paleontologist that mastodonts, like modern elephants, were either "right-tusked" or "left-tusked."[7] Apparently, in those instances where both tusks of the same mastodont are preserved, one is usually longer and larger than the other and shows more wear—evidence that mastodonts used one tusk almost exclusively in prying off and breaking up spruce branches into chewable morsels. If there hasn't been a census on mastodont tuskedness, there should be. Being right-handed, I imagine that most mastodonts favored the right tusk. Lefties, of course, will root for the southtuskers.

# CHAPTER 20

*The babirusa, a wild Indonesian species of pig that almost passed for kosher. (From William Jardine,* The Naturalist's Library. Mammalia, *Vol. V:* Pachyderms, *W. H. Lizars, Edinburgh, 1836)*

# Chewing the Cud

Ever since man met pig he's either sworn at them or by them. Ancient Cretans held pigs sacred because, according to legend, Jupiter was suckled by a sow. Greeks and Romans sacrificed pigs to Hercules, Venus and the Lares (household gods) in the hope of relieving body ailments. Early Egyptians, impressed with porcine fecundity, would slay pigs at royal weddings as a reproductive blessing. They also sacrificed swine to Bacchus and the moon. In today's era of pigskin worship, the people at the University of Arkansas hail their football team as the "Razorbacks," a semiwild hog.

Well, one culture's sacred cow is another's scorn. Muslims and Jews classify pigs as unclean animals and forbidden food, a stricture that dates to between 2700 and 1800 B.C. Consider the uproar then when the news media reported recently that the babirusa, a wild Indonesian species of pig, had been lumped mistakenly with all other unclean swine. According to the press reports, the babirusa, unlike other swine, was really a pig in kosher clothing, in keeping with Jewish and Muslim dietary laws and fit to eat.

The economic implications of this revelation, if true,

would be enormous. Domesticated herds of babirusas would be a cheap, new source of protein for southeast Asian peoples, and would finally provide what observant Muslims and Jews have been missing for so long: kosher bacon, ham, pork chops and babirusas-in-a-blanket.

The "kosher babirusa" rumors grew out of a study by an animal-science professor at the University of Florida, Fuller Bazer. He had been working with the Agency for International Development on a project to increase protein foodstuffs in southeast Asia. Apart from studying fish and poultry production, Bazer entertained the intriguing possibility of domesticating the wild babirusa for human consumption. An anatomical report published in 1940 hinted at critical differences between the babirusa and other swine—differences that might satisfy the Judeo-Muslim dietary laws.

The taboos against pork are rooted in the Old Testament and the Koran. Their dietary laws regarding meat condone only the flesh of animals that both chew their cud (ruminate) and have cloven hoofs. In zoological talk, kosher means the flesh of ruminant artiodactyls. Artiodactyls are an order of mammals that arose 55 million years ago. Among their evolutionary trademarks are an even number of toes (two or four) on each foot, specialized ankle and wrist bones for quick takeoffs, running and climbing hills, and Cuisinart-like cheek teeth for processing bales of vegetation. Next to rodents and bats, artiodactyls are the most diverse among living mammals. They include pigs, peccaries (South American piglike animals), hippos, camels, chevrotains, giraffes, antelope, buffalo, bison, cattle, deer, elk, moose, caribou, sheep, goats, pronghorns.

Most of these animals have two toes—the biblical "cloven hoof"—on each foot, and many are also cud chewers. Exceptions here are the hippos, which have four toes and don't ruminate, and the pigs and peccaries, which only make it halfway to the kosher table: They have cloven hoofs but do not chew their cud. Except perhaps the babirusa. The 1940 anatomical study hinted that the babirusa might be capable of cud chewing and thus fit the kosher bill.

The surest way to tell if an artiodactyl is a ruminant is to watch it eat in its natural habitat. This is difficult if the animal is as secretive and rare as the babirusa. Failing direct observation in the wild, a rough—but not surefire—guide to cud chewers is the number of chambers in its stomach, and it was stomach anatomy that brought the babirusa into the limelight.

Most ruminants have a four-chambered stomach, an evolutionary adaptation for extracting as much nutrient as possible from low-grade plant food such as grass, hay, alfalfa and other forage found on the range and at organic food emporia. By extrapolating from the anatomy of fossil dentitions, paleontologists conclude that the ruminating stomach arose in artiodactyls about 30 million years ago. By that time, a cooler and drier climate had caused much of earth's tropical forests to be replaced by savannas, grasslands and temperate forests north and south of the equatorial zone. These new environments changed the evolutionary ball game for plant eaters. Tougher vegetation favored the survival of herbivores with efficient, long-lasting, grinding dentitions. As a result, lineages of artiodactyls, horses, rodents and other plant eaters evolved a battery of ever-growing, high-crowned cheek teeth for processing coarse roughage. But only among the artiodactyls was the multichambered ruminating stomach also a part of the evolutionary package.

Ruminating allows eating at a leisurely pace, an activity that reaches a high art in Mediterranean countries. Ruminating artiodactyls know the meaning of a languid chew.

They graze and browse huge quantities of plant material, which is swallowed quickly with little chewing. The animal then retires to a safe spot to ruminate. The swallowed food enters the first chamber of the stomach (the rumen or paunch) where it is softened and regurgitated into the mouth. Here the cud is thoroughly chewed and ground, swallowed again, and passed progressively to the second (reticulum), third (omasum) and fourth (abomasum) stomach chambers, where the food is digested through chemical and bacterial action.

The rub for the babirusa is that not all cud chewers have a four-chambered stomach. Evolution graced only deer, elk, moose, caribou, giraffes, pronghorns, antelope, bison, buffalos, sheep, goats and cattle with a gastric quartet. The stomach in camels and chevrotains is three-chambered and also ruminating. In hippos it is three-chambered and nonruminating, and in pigs and peccaries two-chambered and not ruminating. Is the babirusa like other pigs?

The 1940 anatomical study, which was performed on a babirusa that had died in the Chicago Zoo, said no. Its stomach was intermediate in complexity between that of typical pigs and sheep, namely, it had two and one half or three chambers, and might, therefore, be capable of ruminating. More circumstantial evidence for a cud-chewing babirusa was its diet. In the wild, babirusas feed primarily on foliage like other ruminants, whereas most swine prefer tubers, roots and meat and will essentially eat anything. The combination of a foliage diet and a more complicated stomach made the babirusa a possible candidate for membership among bona fide ruminants.

This was evidence enough for the news media. After little rumination, they succumbed to hog-hype and presented the babirusa as a pig with cud-chewing credentials. Bazer, the University of Florida professor, quietly continued his research and confirmed the opposite and standard zoological teaching: The babirusa is not a cud chewer. Although the babirusa's stomach is somewhat three-chambered, it is not capable of rumination. A similar verdict was issued by J. David Bleich, a professor of Jewish law at New York's Yeshiva University. Confirmation also came straight from the babirusa's mouth: At the Los Angeles Zoo there's a babirusa that's never been seen chewing the cud. As it turns out, the matter was actually resolved long ago. Sir Richard Owen, the eminent nineteenth-century English anatomist and paleontologist, described the nonruminating character of the babirusa stomach almost one hundred years before the babirusa died in the Chicago Zoo.[1]

Bleich's ruling, however, added a few new wrinkles. According to Jewish law, Bleich says, the identity of an animal is determined from descent rather than strict anatomy. So far so good—Bleich and Darwin are speaking the same language. But Bleich continues: Because the babirusa, like all living swine, stems from nonruminating (and nonkosher) fossil pigs, it too, by virtue of its ancestry, is not kosher. This rule would apply even if the babirusa had independently evolved a cud-chewing capability. Bleich's opinion neatly precludes the "creation" of a kosher animal from a nonkosher one through genetic engineering.

However, in a strictly evolutionary sense, Bleich's ruling would also render *all cud chewers* nonkosher, because ultimately they too have a nonkosher pedigree. Sheep, goats, cattle and all other artiodactyl ruminants originated millions of years ago from nonruminating forebears, just as did the babirusa.

If there's a lesson here, it's that there's more to chewing the cud than counting stomachs. And there's more to the kosher laws than cloven hoofs and chewing the cud. After all, for five thousand years Near Eastern Semitic peoples thought that pigs were good to eat. What made them change their minds and enact the taboos?

One common suggestion is disease. To paraphrase Ambrose Bierce, trichinosis is the pig's reply to pork eaters. The trouble with this idea is that the risk of disease did not keep pork from being a dietary mainstay before the taboos were installed.

A second theory blames the Indo-Aryan nomads, who spread new ideas and technologies throughout eastern Europe and western Asia.[2] These nomadic pastoralists tended goats, sheep and cattle but loathed pigs, perhaps because swine lack the stamina and temperament for open-country herding. The Indo-Aryan antiswine doctrine may have reached the Near East at about the time (1800 B.C.) some authorities date the onset of the dietary laws.

Maybe so, but this scenario fails the ecological test. The Indo-Aryan prejudice arose from herding livestock on grasslands, a habitat unsuitable for pigs. Swine are forest browsers and it was in forested areas (or open sties) that the settled inhabitants of the Near East happily tended their domesticated pigs. They would have little reason to accept a bad hog rap.

One major anthropological school holds that the dietary laws and most other societal strictures were "cultural shields" erected to safeguard cultural unity, namely, to prevent assimilation and intermarriage. According to this viewpoint, a supplementary reason for defining swine as unclean may have been to discourage idolatrous practices, such as pig worship and sacrifice. Unfortunately, both ideas seem to be putting the cart before the swine. Cultural shields aren't usually legislated or predesigned; they develop as by-products of different practices. For example, I wouldn't normally marry a woman from a culture that regularly feasted on cockroaches. But the cockroach cultural shield wasn't dreamed up to prevent our marriage.

A brutally pragmatic line of reasoning to explain the dietary laws comes from a school of anthropology that calls itself "cultural materialism." It says that peoples have developed most of their sociocultural behaviors in response to ecologic or economic need. According to Marvin Harris, a noted exponent of cultural materialism, the pork taboo grew out of ecologic change and pig-human competition for the same food resources.[3] Pigs are adapted for browsing in forests. In the Near East, intensive farming and herding activities from 7000 B.C. to 2000 B.C. led to a drastic deforestation, erosion and drying of the land, and a severe shortage of areas suitable for raising pigs on natural forage. As a result, Harris reckons, herds of domesticated pigs had to be fed large quantities of grain, which up to then had been grown mostly for human consumption. This proved to be too great a siphon on human food resources; the cost (in grain foodstuffs) of raising pigs greatly outweighed the return in pork meat and fat. The taboo against swine served to stem a long-term economic disaster. As a religious doctrine, it ensured compliance, especially by hog-minded farmers.

Harris's thesis, although intriguing, begs some questions. A sacred taboo would work, but wouldn't the extermination of all pigs in the region work much better? How is the alleged cost of pig-human competition reconciled with widespread pig husbandry in areas ecologically similar to the Near East, such as central and northern Asia? At least, Harris's explanations can be tested scientifically, unlike some of the others.

None of these theories have much to do with the babirusa. It was never domesticated and now probably never will be, seeing that it can't pass for kosher. All I can say is, I'm glad I didn't rush off to the brokers at the Chicago Hog Market and put my money into babirusa futures.

# CHAPTER 21

*The faces of man: physiological visages or physiognomy verité? (From Oliver Goldsmith,* A History of the Earth and Animated Nature, *Blackie & Son, Glasgow, 1840)*

# Face Value

Emotions used to be simple. Time was I'd get up in the morning, face the mirror and grimace. So would you if you came face to face with my face: unkempt, skin like a moonscape, eyes like drifting continents. I always thought my grimace was straightforward cause and effect. Emotional stimulus: face like a stomachache. Facial response: grimace.

Apparently not. It seems I might grimace whether I had a face like Clark Gable or a Clark Bar, at least according to a school of psychology that is pulling an about-face on emotion and facial expression. Common sense (and Darwin) always had it that facial gestures evolved to broadcast our emotions. But some psychologists think it works the other way around: Facial expressions precede and cause the emotion.

This idea first surfaced in 1907 when Israel Waynbaum, a French physician, published a radical theory of emotional expression.[1] Waynbaum held that facial gestures involve the contraction and relaxation of facial muscles, which constrict and dilate facial blood vessels and thereby regulate the blood flow to and from the brain. This, in turn, stimulates emotion. In other words, I don't necessarily grimace in the morning because my face could stop a clock. No. My grimacing muscles pinch certain facial blood vessels, affect the blood supply to my brain and induce the blues.

Let's take a less personal example. Normally, we think we smile when we feel good because a good feeling made us smile—a principle that governs used-car salesmen. Not so, says a smiling Waynbaum. He claims that the physical act of smiling—the contraction of cheek muscles—constricts facial blood vessels and impedes the flow of blood in the face, causing a temporary overabundance of blood (and oxygen) around the brain and a "good" feeling. Similarly, grimacing, frowning, pouting and the like will cause temporary brain anemia and feelings of depression. So, in effect, next time you smile or frown, get ready to face the feeling on your face.

This theory reminds me of my father, who had his own notion of the causes of facial demeanor. He subscribed to the gaseous theory of emotional expression: Facial gestures were ultimately due to the state of one's digestive vapors. My sour visage meant I had too much gas. My happy countenance signaled gaseous equilibrium. It only goes to show that one doesn't have to be French or a physician to practice naïve empiricism.

Soon after it was published, Waynbaum's theory of vascular control of emotion was discredited on physiological grounds. Nevertheless, it recently received a face-lift from R. B. Zajonc, a psychology professor at the University of Michigan.[2] Although Zajonc reclothed Waynbaum's theory in the skin of modern anatomy and physiology, the basic premise is still the same: Facial gestures, through the contraction and relaxation of facial muscles, mediate cerebral blood flow and affect emotional states. Zajonc's new physiological wrinkles involve the influence of cerebral blood flow on brain temperature, on the release and action of neurochemicals and, in turn, on subjective emotions.

The evidence cited by Waynbaum and Zajonc is mostly rhetorical, and, as we all know, rhetoric is gas in a three-piece suit. Why, asks Zajonc, should humans have evolved such a complex repertoire of facial gestures just to express emotions? Isn't that the role of language, the hallmark of our evolutionary history? Therefore, some other adaptive function must be at the heart of facial gestures, namely, the internal regulation of cerebral blood flow, brain temperature, neurological activity and, as a result, emotional states.

Furthermore, why are many facial gestures—say, furrowing of the forehead—symbols common to all cultures? Waynbaum and Zajonc argue that the real common denominator is physiology, not behavior or communication. For example, intense thought is associated with furrowing the forehead and projecting the lower jaw in almost all human cultures. The contraction of specific muscles involved in these gestures momentarily shunts more blood from the face to the brain. And for the brain, more blood means more "thought." Zajonc also invokes facial control of cerebral blood flow to explain the universality among humans of laughing, smiling, crying, lip-licking, frowning and blushing.

At first blush, this vascular theory of facial gestures makes some mechanistic sense. True, compared to other primates, humans have a wealth of facial expressions. We use about eighty facial muscles to grimace, glower, glare, leer, sneer, frown, furrow, scowl, sulk, smirk, blush, flush, laugh, smile, grin, wink, beam, yawn, flash/lower/raise an eyebrow, bare teeth, look blue, look black and keep a straight face. These expressions in all cultures signal the nuances of human sentiments.

But perhaps Zajonc should have furrowed his brow a bit longer and stuck out his jaw a bit farther before reviving Waynbaum's ideas. For one thing, modern physiology flies in the face of their theory. Critics point out that there are no solid physiological bases to link facial gestures with control of blood flow to and from the brain.[3] Blood reaches and leaves the brain through a profusion of arteries and veins, too many to to be regulated significantly by facial muscles. Also, patients with facial paralysis don't suffer impaired emotional reaction. Moreover, experiments indicate that more blood does not a thinking brain make; nor a smile, nor an emotional high. Finally, blood flow to and from the brain is too slow to account for the almost instantaneous onset of emotional states. If there is a link between facial gestures and emotion, it would probably involve neural feedback from the contracting facial muscles. During smiling, for example, the nerve impulses generated by the contracting cheek muscles may reinforce the emotion that originally caused the smile.

There is more than just physiological counterevidence. History adds the monocle as a good test of the Waynbaum/Zajonc theory. Monocles were popular during the eighteenth century, either as a fashion accessory or for corrective vision. To keep a monocle screwed to the eyeball the wearer had to scrunch up half his face, namely, contract the facial muscles that raise the cheek and one corner of the mouth. Trouble is, not only does this keep the eyepiece in place, it also contorts the monocled half of the face into a perpetual, grotesque smile. If Waynbaum and Zajonc are right, the monocled should have had a monopoly on smile-induced cheerfulness. But that's not how historians see the Prussian military elite, for whom the monocle might as well have been a part of the uniform. The Prussian general staff, despite their monocled, smiling half-faces, never managed to pass for the good-humor boys.

Charles Darwin faced the origin of emotional expressions thirty-five years before Waynbaum unveiled his vascular theory. In 1872, Darwin published *The Expression of the Emotions in Man and the Animals,* in which he made a comparative survey of facial gestures among primates and other mammals to uncover the roots of human expression. To Darwin, behavioral features in animals were no different from anatomical ones. Their origin and nature were due to evolution by natural selection, because they were as critical to survival as physical traits. Postures, gestures and facial expressions had arisen, been sorted, modified, fixed and genetically transmitted among species over millions of years of evolutionary descent.

Ethology, the modern discipline of animal behavior, has since corroborated and extended Darwin's synthesis. Specific behavior patterns, like anatomical traits, can confer a natural advantage in the competition for mates, space and food; such behavior patterns will be inherited more frequently in each generation until they become prevalent in a species.

The adaptive value of facial expressions is communication, especially among social organisms, where more efficient communication may increase the chances for survival and reproduction. The evidence culled by Darwin and others is compelling. It indicates that facial gestures signal intent—the mood and probable behavior of the gesturing animal—and that many human facial expressions are exhibited in one form or another by primates and other mammals. Smiling is one of these. Many mammals flash a "bared-teeth display" to signal threat, or fear, or the intent to flee. In some apes and monkeys the same baring of teeth has evolved into a "fear-grin" or "nervous smile," an apparent gesture of submission and reassurance that during a confrontation reduces the chances of fighting and flight.

Among humans the meaning of the smile signal is similar, ranging from a sign of caring (remember Jimmy Carter's toothsome face?) to confidence (John Kennedy) to "father knows best" (Reagan) to cunning coexistence (Khrushchev at the U.N.). The human smile is also innate. It appears between two and four months in infants in all cultures and, significantly, even in deaf and deaf-blind children who were never taught to smile.

Smiling and other common facial gestures that signal across cultural lines are part of the evolutionary baggage inherited and modified by *Homo sapiens.* Originally, these behaviors evolved from our hominid and earlier primate ancestors as nonverbal gestures to reflect emotion and intention. But now most facial gestures are not silent, isolated signals. They go cheek to jowl with speech, or, as Darwin put it, they "give vividness and energy to our spoken words." More than two thirds of our facial expressions are paralinguistic—they accompany and reinforce language.

For many investigators this is good evidence that most gestural communication—hand, arm and facial signals—preceded and then coevolved with spoken language in our immediate ancestors, *Homo erectus* and early *Homo sapiens.* Other anthropologists and linguists disagree. They think that language could have arisen by other means: the imitation of animal sounds; the imitation of sounds produced by objects being struck; primitive work grunts; mouth gestures that mimic body movements; spontaneous infant babbling about the external environment; primitive grunts of pain, pleasure, fear, surprise and the like. Critics, in turn, have poo-pooed these ideas by nicknaming them the bow-wow, ding-dong, yo-he-ho, ta-ta, babbleluck and ouch-ouch theses of language origin, respectively.[4]

Older notions of the origin of language are no better. In the sixteenth century, King James IV of Scotland ordered two infants raised without exposure to language, on the premise that when the children finally spoke it would be in man's primeval tongue. According to the king, their first babbles were in Hebrew, in keeping with the language ascribed to Adam and Eve by biblical scholars of the day. More than a hundred years later, a Swedish scholar disagreed. He concluded that in the Garden of Eden God had spoken Swedish, Adam, Danish and the snake, French.[5] Some garden. Sounds more like a meeting of the European Common Market.

Whatever the origin of language, if Waynbaum and Zajonc are right about facial expressions, it would put a different face on art, literature and history. For example, was it only physiology behind the face that launched a thousand ships? Did the Hapsburg monarchs, famous for their projecting lower jaws, spend much more time in intense thought than history records? And was there nothing more to the Mona Lisa's enigmatic smile than a pretty gal with too much blood on the brain? Of course, if the recent suggestion is true that the Mona Lisa is really Leonardo da Vinci's self-portrait in drag, then we can't take her or his smile at face value. He may have had more than just blood on the brain on his mind.

# Chapter 22

*Macacus cynomolgus*, Javanese macaques.

*Monkeys, the animals most scorned by creationists, share a 70-million-year evolutionary kinship with all other primates: humans, apes, lemurs, lorises and tarsiers. (From John S. Kingsley,* The Standard Natural History, *Vol. 5:* Mammals, *S. E. Cassino and Co., Boston, 1884)*

# Monkey Trials

Charles Darwin would be saddened to learn that on May 5, 1986, more than a century after he published *Origin of Species,* the Supreme Court of the United States entered the battle over teaching creationism as science in the public schools. The issue facing the Supreme Court hasn't changed since Darwin's time or, for that matter, since *Homo sapiens* began to wonder how the universe ticks.

The case[1] grew out of Louisiana's "Balanced Treatment for Creation-Science and Evolution-Science in Public School Instruction" Act. It was passed by the state legislature in 1981 and mandated the teaching of "creation-science" whenever evolution is taught. The legislation was struck down as unconstitutional by a U.S. district court in Louisiana on the grounds that the act promoted a religious belief ("creation-science") and violated the establishment clause of the First Amendment—the separation of church and state. When the district court's decision was subsequently upheld by the U.S. Court of Appeals for the Fifth Circuit, Louisiana appealed to the Supreme Court, which, on May 5, 1986, agreed to hear the case. Ironically, sixty-one years before, on May 5, 1925, John Scopes had been arrested in Tennessee for teaching evolution.

The Louisiana creationism case is an instant replay of an Arkansas case. On January 5, 1982, a U.S. district judge, William R. Overton, overturned an Arkansas law that required "balanced" treatment for evolution and "creation-science" in its public classrooms. His conclusion that creationism "has no scientific merit or educational value" came after a sad, latter-day resurrection of the 1925 Tennessee Scopes trial. Plaintiffs against the Arkansas law included clergymen from six Christian denominations, three Jewish organizations, educators, parents of local schoolchildren and the American Civil Liberties Union. Despite the verdict, the Arkansas statute begat the Louisiana act.

Perhaps now the Supreme Court finally will do for Darwin what was done long ago for Copernicus, Kepler, Galileo and the solar system: rule that the discovery of natural laws governing the universe and earth may not be suppressed by nor pitted against particular religious dogmas. Natural laws also govern life on earth, its origin, history, change and extinction through geologic time. The story is grand, the knowledge dazzling. Both have been acquired and honed by scientific inquiry. Creationists would replace inquiry with mythological pap during a full-scale retreat to a pre-Copernican world.

These invitations to return to a simplistic past are not new. During the sixteenth and seventeenth centuries humankind and religious creed had to face the scientific music, the discovery of the natural, physical laws of the universe: The earth, and we on it, were not the center of the solar system. Neither were we select among the heavenly bodies. Earth merely orbited the sun along with other planets. Galileo nearly paid with his life for that piece of music.

Ever since Darwin, mankind has faced a second comeuppance: We are not the product of a divine, instantaneous creation. Rather, our origins are as humblingly genetic and geologic as those of all other organisms. Also, the diversity and distribution of fossil and living organisms on earth did not arise during one "creation week." Rather, they are rooted in over three billion years of evolutionary history. As was the case with the Copernican universe, the fact of the evolution of life on earth is perceived as a threat by a particular fundamentalist minority. Evolution contradicts that minority's strictly literal reading of their scripture and their strictly scriptural reading of human origins.

The facts facing the Supreme Court are clear. "Creation-science" is not science. It cannot predict or explain phenomena in the natural world. Its tenets cannot be tested or falsified by observation or experiment. It shuns naturalistic causes in favor of supernatural ones, comprehension of which is inaccessible to us. It systematically ignores, misunderstands and misrepresents a massive body of knowledge of the universe in favor of a particular religious revelation—a literal, word for word reading of Genesis—which, unlike scientific hypotheses, cannot be amended or abandoned.

Although the Louisiana act defines "creation-science" as "the scientific evidences for creation and inferences from those scientific evidences," this is a semantic game. There are no such "evidences." Instead, "creation-science" proclaims what is still unknown about the evolution of life on earth as evidence for creation. In so doing its proponents commit the kindergarten blunder of mistaking the absence of evidence for evidence of absence. Alleged "evidences" of "creation-science"—such as the coexistence of human and dinosaur footprints in the Paluxy riverbeds of Texas, or a young earth indicated by the rate of decay of the magnetic field—are now admitted by the Institute of Creation Research to be fraudulent or simplistically wrong. Indeed, the word "evidences" in the Louisiana act belies the "science" in "creation-science" and reveals the religious intent of the legislation. "Evidences" is a curious word found only in the Dictionary of Christian Apologetics. Science uses "evidence."

Evolution is not the sole target of the creationist assault. "Creation-science" would erase most of science, art and the humanities. The singling out of evolution for censorship or for "balanced treatment" with creationism in the public classroom begs a larger question: Will we dismiss all knowledge? For example, the "creation-science" notion of a universe no older than 10,000 years would banish all of astronomy, physics and planetary geology. Its belief in an earth no older than 10,000 years would deny all of geology, paleontology, archaeology, biology, biochemistry and chemistry. In essence, creationists would have it that the rational world has been conned by the fossils and rocks, flimflammed by the bodies and light of the universe, duped by the nature of living organisms and deceived by our senses.

Evolution is but a part of cosmology, the science of the universe and its systems, including life on earth. It is as fundamental a cosmological player as the theories of gravitation, relativity, planetary motion, radioactivity, plate tectonics and light. To teach "creation-science" as science means to plunge our classrooms into the dark, and relight the candles of a superstitious medieval world.

The notion of "balanced treatment" is meant to appeal to our sense of fair play, which might apply to balls and strikes, but not to willful ignorance or archaicism. For that reason we don't give equal time and credence in geography or astronomy courses to the ideas of a flat earth or an earth-centered solar system. (Unbelievable as it may seem, there are "creation-science" extremists who believe in the latter theory as a scientific tenet blessed by scripture and maintain that geocentrism as well as creationism should be taught in the classrooms.) Equal time implies equal substance. "Creation-science"—a fundamentalist, literal Genesis-reading of origins—should not be allowed to bully its way into science classes and texts under the flummery of fair play or the facade of science.

Even if reason prevails in the courtroom, it may not in the schoolroom. Too many state and local school boards have succumbed to creationists' pressure to adopt biology textbooks that either soft pedal evolution or give equal space and credence to "creation-science." And textbook publishers, with a greater interest in book sales than in scientific integrity, are all too willing to comply in excising "offensive" evolutionary statements from biology texts.

Perhaps most unfortunate is the false apposition of science and religion by proponents of "creation-science." Reduction of a complex issue to a simplistic dualism is a favorite trick of demagogues; here it devalues the currency of both science and religion. To the sponsors of the Louisiana act, evolution is just the proverbial tip of the Satanic iceberg: "evolution-science" is anti-God and "creation-science" pro-God, period. Ardent creationists place the sciences and humanities at the root of all evil and "non-creationism" at the foot of immorality. Strange, then, that the plaintiffs in the Arkansas and Louisiana cases against the teaching of "creation-science" included Catholic, Protestant and Jewish clergy, as well as teachers, educational organizations, parents and schoolchildren.

Indeed, many religions have no trouble accommodating belief in a Creator with the facts of evolution, gravity and motion, a sun-centered solar system 4.6 billion years old, and an ancient universe 15 billion years old. A conflict between science and religion will be forced, as it is by ardent fundamentalists, only if chapter and verse of scripture are read in dogma, not spirit.

If proponents of "creation-science" studied the history of evolutionary thought as fervently as they do scripture, they would learn that it was the great natural theologians who paved the evolutionary road for Darwin. They cataloged, classified and ordered nature, organic and inorganic, living and fossil. To them Nature was God's handiwork, and in it they perceived the grandeur of creation manifest in the forms of life and their rhythms of change. They anchored God at the beginning of an evolved nature as the ultimate progenitor; once Nature had been set on its evolutionary way, He did not dabble with the coming billions of anatomical mutations and transformations. Otherwise, as one of them, Robert Chambers, asked, would God really spend His time invoking special miracles just to alter the tubercles on teeth of different species of mice? The natural theologians saw in evolution a divinity more exalted, a deity more personal and an origin more magisterial. If creationists would read *Origin of Species*, they would learn that Darwin saw likewise.

## Epilogue

My vote against teaching the King James version of the universe as science brought a rush of mail, half of which favored creationism. One reader went so far as to cancel his membership in Carnegie Institute. Another offered me a $5,000 reward if I could "show that evolution is scientifically possible." I declined, considering his ground rules and the slim odds that any proof would be acceptable. It would have been easier to teach a pig to play the flute or a babirusa to chew its cud.

The same principle holds for public face-offs between

"creation-scientists" and evolutionary biologists: They only convert the converted. One of these debates was scheduled for August 1986, during the International Conference on Creationism in Pittsburgh. The conference title was "The Age of the Earth." I could have saved the conferees a heap of time, trouble, travel and expense: By all evidence and observation, the earth is approximately 4.6 billion years old. They could have saved their speeches, stayed home and enjoyed our antiquity.

It's too much to hope for. Consider that at a recent Bible Science Association conference in Cleveland, the climax of the program was an evening debate over our solar system: Is the sun at the center or is the earth? Hell, why not put Uranus at the center and be done with it. They also had trouble finding champions of a sun-centered solar system among the conferees. Maybe I can get a leg up on next year's conference by claiming that the earth is round.

On December 10, 1986, the Supreme Court heard the appeal from the State of Louisiana. Six months later, on June 19, 1987, the justices agreed 7 to 2 with the lower courts that the Balanced Treatment Act sprang from fundamentalist religious fervor and that "creation-science" intended to advance a particular religious belief. The most discouraging part of this episode in judicial history is that the High Court ruling wasn't unanimous. The two dissenters, Chief Justice Rehnquist and Justice Scalia, citing a perverse definition of academic freedom, chose to don blinders so as not to see religious dogma dressed in transparent scientific clothing.

# CHAPTER 23

*A geological eruption on Iceland. By the All Star break, Iceland and Hawaii will have suffered many more eruptions, and Billy Martin and the Yankees will have parted one more time. (From Louis Figuier,* The World Before the Deluge, *D. Appleton and Co., New York, 1867)*

# Natural Predictions

Predictions are a risky business. After all, in 1979, a famous "psychic" foretold that the ERA Amendment to the U.S. Constitution would be passed into law, the Russians would launch a military attack on Israel and the Ayatollah Khomeini would die after being pushed off a mosque in Mecca. So much for that psychic. The ayatollah is still with us, ERA isn't and the Russians knew better than to fool with a country with 3 million people and 102 political parties. They went after Afghanistan instead.

Then there is the popular prophet who did not profit from her prediction that Russia would be the first to put a man on the moon, or that St. Helen's Island, the site of Montreal's 1967 World's Fair, would unhinge and float down the St. Lawrence River sometime during EXPO 67. It still hasn't, twenty-two years later. The psychics were skunked again in 1985 when they divined that Prince Charles would be trampled by elephants, Lee Majors would go into space and—the perennial favorite—alien beings would make contact with Earth.

Despite these foreboding omens, I offered my services as a prognosticator to the supermarket tabloids at the beginning of 1983 by making seven shocking predictions for that year.[1] I thought I could do no worse than that talented oracle, former Yankees and Mets manager Casey Stengel. When asked about the prospects of a certain rookie ballplayer, Stengel predicted: "He's twenty-two now, and in ten years he has a good chance of becoming thirty-two." In that spirit, I ventured the following seven astounding predictions on January 1, 1983:

1. Ape-man 1: "Bigfoot," the hairy giant "ape-man" of the Pacific Northwest, will suddenly reappear.

2. Ape-man 2: Expeditions in East Africa and/or Asia will unearth the spectacular remains of an extinct "ape-man" millions of years old.

3. Physicists will discover a new, mysterious subatomic particle.

4. Devastating earthquakes will rock South America, Asia and North America.

5. The Himalayan Mountains will rise even higher.

6. Fiery volcanoes will erupt on Hawaii and Iceland.

7. The entire North American continent will be jarred west from where it is now.

My clairvoyant powers proved auspicious. Only six months later, by July of that year, predictions #1 and #3 to #7 had already come to pass.[2] My prophetic average was 6 for 7, I was batting .856 and was ready for the call from the supermarket tabloids with an offer of front-page billing. None came.

Five years later, times have changed. The prognosticating market is no longer cornered by the tabloids. Their 65-cent seers are out of fashion and their prophetic tools passé, except, of course, in the Reagan White House. Supermarket swamis are out, boutique prophets are in. Now the audience is upscale, the prices fancy and the oracles a clique of designer diviners: New Age "astral travelers" and "trance channelers."[3]

Well, New Age or old, there's never been a shortage of dupes hungry for hokum. And upscale or not, this audience can still be identified by its lazy intellect and willful abandonment of reason. The hokum hasn't changed: A quick dose of simplistic pablum sold at the price of complex inquiry.

Nonetheless, I offer the same scientific, time-tested geomantic predictions for 1988, 1989 and beyond that I first proffered in 1983. I'm glad I have science on my side. Everyone knows that science can predict an eclipse of the sun years in advance but can't forecast a blowout over the weekend.

Prediction 1: "Bigfoot" sightings

The existence of an upright-walking "Bigfoot" creature in the Pacific Northwest has been claimed for many years. Alleged sightings culminated in a famous Bigfoot film taken at Bluff Creek, California. More evidence of this two-legged Godzilla took the form of giant seventeen-inch footprints discovered in Oregon, Washington and northern California. Unfortunately, the Bigfoot creatures turned out to be bears standing on two legs, the film turned out to be either of a bear or a man in a bear suit on two legs and the footprints turned out to be a two-legged hoax.

The ruse started in 1930 when Rent Mullins, a longtime resident of Toledo, Washington, whittled a pair of nine- by seventeen-inch wooden "feet" and began stomping impressions in forest floors, along lake shores and in people's minds. In the following years, Mullins became a Bigfoot foot distributor, making eight sets of wooden monster feet for friends and hoaxers in California, Washington and Oregon. By January 1983, Mr. Mullins's practical joke was up, admitted and well publicized. But, true to my prediction, an April 1983 issue of a supermarket tabloid carried the following sensational headline:

BIGFOOT STOLE MY WIFE
DISTRAUGHT HUBBY VOWS REVENGE

If this is true, all I can say is that his wife would have been a bombshell in the Pleistocene. In a second incident, more purported footprints of this apelike beast surfaced in Oregon in April 1983.[4] The new prints were spotted after a forest ranger reported "seeing a hairy, erect animal unlike a bear." I don't know what all the fuss is about. Lots of places are full of hairy, erect animals unlike bears. Bulgaria. Soviet embassies. The south of France. Quebec's Legislature.

Unlike Mr. Mullins's footmade prints of Bigfoot, the Oregon prints purportedly have thousands of fingerprint-like squiggles, apparently like those of the Tibetan "abominable snowman" or yeti. This, believers say, make the yeti and Bigfoot birds of a footprint or primates of a feather or whatever metaphorically mixed kin one chooses. Essentially, the prints are used to buttress the notion that a large, bipedal, apelike primate "evolved in China five million years ago and migrated [across the Bering land bridge] to the Pacific Northwest"[4] where it now reigns as Bigfoot.

The discovery of new living species, usually new insects and plants, is common, especially in areas of the world that naturalists have barely explored, like the equatorial regions. The same applies to the discovery of mythical beasts: Contact with new cultures uncovers new animal myths. For instance, Pygmy tribesmen of the Congo "believe" in the existence of a dinosaurlike beast they call Mokele-Mbembe. An expedition to the area by a University of Chicago scientist, Roy Mackal, confirmed the myth but not the dinosaur. Similarly, the yeti "lives" in the mythology of the Sherpas of Tibet, as does the Sasquatch (or Bigfoot) for certain tribes of Native Americans. Myth becomes reality only when the evidence becomes unequivocal: The body of the beast is brought back, dead or alive. Otherwise, only wishful thinking, at best, or willful fraud, at worst, will animate a blurred photograph or squiggly footprint.

For example, all attempts to confirm the existence of a plesiosaurlike monster in the murky waters of Loch Ness, Scotland, have failed, beginning with Saint Columba's "sighting" of "Nessie" in the sixth century and ending with the recent plumbing of the loch by sonar and submersibles. Too bad. No one would be more delighted than a paleontologist at the prospect of a firsthand look at a prehistoric reptilian "sea-monster," a plesiosaur that by all geological accounts became extinct with the dinosaurs over 65 million years ago at the end of the Age of Reptiles. Still, Nessie breathes and beckons in tourist brochures for the Scottish highlands. To ensure a glimpse of Nessie, my advice is to fly to Scotland, take the train to the highlands, do the malt whiskey tour a couple of times and hurry quickly to the banks of Loch Ness.

### Prediction 2: Discovery of fossil "ape-men" in East Africa and/or Asia

Fossil "ape-men" or, more correctly, fossil hominids (the family of humans and prehumans) have been recovered with regularity over the past three decades from East African and Asian sites dating between 1 and 12 million years ago.[5] In 1964, Louis and Mary Leakey found parts of the skull and skeleton of a new, extinct species of *Homo* (the genus of modern man) in 1.8-million-year-old sediments in Olduvai Gorge, Tanzania. They named it *Homo habilis* ("handy man"), after the humanlike fossil hand bones that had crafted our earliest known technology—an array of primitive stone tools found in the same deposits. Five years earlier, Olduvai Gorge had yielded the skull of another new "ape-man," *Zinjanthropus boisei* (nicknamed "nutcracker man" after its enormous teeth), a species that has since been placed in the genus *Australopithecus* ("southern ape-man").

In 1972, an expedition to East Turkana in northern Kenya led by the Leakeys' son Richard unearthed the famous "1470" specimen, the first nearly complete skull and face of *Homo habilis.* Two years later, Donald Johanson and a team of American, French and Ethiopian scientists stunned the paleoanthropological world with news of finding the oldest, most complete skeleton of a hominid at Hadar, Ethiopia. The three-and-a-half-foot skeleton, more than 50 percent complete and 3.5 million years old, was informally dubbed "Lucy," an upright-walking female with a chimpanzee-size brain.

Formally, she and other remains of her kin were placed in a new, distinct hominid species, *Australopithecus afarensis* ("southern ape-man from Afar"), by Johanson, Berkeley paleoanthropologist Tim White and French paleontologist Yves Coppens. Two of their conclusions earned Mary and Richard Leakey's scientific disdain: first, that *A. afarensis* was the direct ancestor of *Homo habilis* (and, ultimately, modern man); and second, that fossil hominid bones from Laetoli, Mary Leakey's site in Tanzania, also belonged in *A. afarensis.*

In 1976, it was Mary Leakey's turn to bowl over the paleoanthropological community. Working at Laetoli, she and her colleagues uncovered a series of ancient volcanic ash beds laid down by successive eruptions of a nearby volcano between 3.5 and 3.8 million years ago. One of these ash falls records only a few hours of prehistoric time during which three hominids, perhaps a male, female and child, strolled across Laetoli, leaving two parallel trails of their footprints etched in the soft, damp ash. Other animals, from insects to gazelles to elephants, also left their prints in the ash layers, almost ten thousand prints in all. Whether the Laetoli hominid was *A. afarensis* or not (most recent interpretations say no),[6] Mary Leakey's discovery was dramatic confirmation that upright posture in the human evolutionary lineage had appeared at least by 3.5 million years ago, 2 million years before the evolution of a large brain in *Homo habilis.*

In 1981, the antiquity of the human upright, bipedal gait was pushed back half a million years. Tim White and I found part of a fossil thigh bone and partial skull of *A. afarensis* in rocks more than 4 million years old in the Middle Awash area of central Ethiopia, just south of Hadar. Three years later, Richard Leakey's expedition unearthed a spectacular hominid specimen on the western side of Kenya's Lake Turkana: almost the entire skeleton of a boy belonging to *Homo erectus,* the evolutionary descendant of *Homo habilis* and the ancestor of modern man. Its bones had eroded out from 1.5-million-year-old rocks near a dry wash only a few hundred yards from Leakey's field camp.

In 1986, a few miles south of this wash, Alan Walker, an anatomist at Johns Hopkins and a member of Leakey's research team, found the skull of another hominid. It was field-tagged as the "black skull," but in scientific print it was *Australopithecus boisei,* the same species Louis and

Mary Leakey had first recognized at Olduvai. At 2.5 million years old, however, the "black skull" is the oldest remains attributed to this species.

These were headline-making discoveries in the public press, but many other critical, though less spectacular, fossil hominid finds in Africa and Asia went unreported in the media. Instead, news of the bones and teeth was confined to technical descriptions in scientific journals. As of this writing, a number of expeditions are out combing the badlands of East Africa and Asia for more skulls and skeletons of our fossil ancestors. One of them will strike hominid pay dirt in 1988.

### Prediction 3: Physicists will discover a new subatomic particle

Another good bet is the one I place on physicists' adding a new subatomic particle to the long list of particular names—an "addon" to muon, gluon, meson, lepton and quark.[7] Back in the prehistoric days of physics and chemistry, protons, neutrons and electrons were easily understood as the basic building blocks of atoms. But physicists began discovering new, more fundamental particles as they learned to build and fire huge linear and ring accelerators, the Daytona Speedways of atoms. When speeding protons and neutrons were raised to ever higher energy states in these accelerators, they were found to consist of smaller, basic particles called hadrons. Hadrons, in turn, were made up of yet more elementary particles called quarks.

Quarks came in three flavors, which physicists named "chocolate," "vanilla" and "strawberry" in an attempt to demonstrate that people in lab coats have a sense of humor. The formal flavors for quarks are "up" quark, "down" quark and "strange" quark. Each quark also has its antiquark (antimatter) partner. Whatever the flavor, quarks have one quirk: They elude direct observation, yet their existence is implied in mathematical equations and in the chambers of particle accelerators. In 1974, a fourth flavor—the "charmed" quark—was found, and in 1978, a fifth—the "bottom" quark. Physicists prophesied the existence of a sixth quark and were so confident of their forecast that they prenamed it the "top" quark. When physicists talk, I listen. In 1983 I predicted that their prediction was correct.

Sure enough, it took only twenty-six days for this prophecy to materialize. On January 26, 1983, Swiss physicists announced the discovery of not one but two new particles that are short in life but long in importance: the W-plus particle, and its antimatter twin, the W-minus. They are fundamental to the workings of the weak nuclear force—one of the four basic forces of the universe. The other three forces are: the electromagnetic force, which is responsible for radiation, such as visible light and X rays and is carried by photons, the unit packages of light; the strong nuclear force, which binds protons and neutrons in the nucleus of an atom through the action of particles called gluons—the atomic version of Velcro; and gravity, which is carried by gravitons, theoretical particles that have not yet been observed but, I predict, will be.

The W particles were created in a big-bang explosion in a huge particle ring accelerator outside Geneva. Two beams, one of protons, the other of antiprotons, were launched toward one another and accelerated to tremendous velocities. In the ensuing collision, the protons and antiprotons were annihilated but their matter was converted into enough energy to create the W particles, eighty to ninety times more massive than protons. Physicists say that this relationship between protons and W particles confirms a twenty-year prediction that the weak nuclear force and electromagnetic force are fundamentally related.

Just as confusing as protons and neutrons is the electron, the third atomic particle I used to understand. It is actually only one of a family of particles called leptons. Other leptons are muons, tauons and neutrinos. Each has a specific mass and each appears to be truly fundamental rather than a composite of smaller particles. Physicists also predict the discovery of more massive leptons when more massive accelerators are built.

### Predictions 4–7: Geologic chaos

These four predictions require a setting of the geologic stage, a short geomantic preamble. Ever since early explorers mapped the earth's oceans, continents and coastlines, suspicions have arisen that continents were not anchored in place, but had somehow sailed across the globe to their present positions. The opposite coasts of the Atlantic Ocean were mirror images of each other. Remove the Atlantic, and North and South America appeared to fit against Europe and Africa; so did India and Madagascar to Africa, Australia and Antarctica to each other and both to Africa and South America. Was it just geographical coincidence that the earth's continents resembled huge chunks of a geological jigsaw puzzle that could be pieced together into a gigantic single landmass, a supercontinent?

No, said a German meteorologist, Alfred Wegener, in 1915, and he predicted geology would prove this radical idea of continental drift. He was greeted with choruses of scientific cheers and boos. The cheerers applauded the simple elegance of the theory and its unified explanation of different geological observations. The booers decried the lack of a known physical mechanism that could propel continental masses about the surface of the globe. Two camps developed, one on either side of the fence of continental drift: the pro-drifters and the anti-drifters.

Pro-drifters marshaled their evidence: rock strata on opposite sides of the Atlantic in Africa and South America were laid down in identical sequences and preserved the same fossil animals and plants. These sediments and fossils dated to geologic periods called the Permian and Triassic, between 200 and 300 million years ago, the time when all of the earth's landmasses were welded into an ancient supercontinent, Pangaea (meaning "all lands"), and surrounded by a single world ocean, Panthalassa ("all seas"). Anti-drifters, although impressed with the fit of geography, geologic strata and fossils, wanted more concrete evidence before jumping the fence—evidence that the earth's crust was dynamic and mobile rather than solid and fixed.

That evidence arrived after World War II from three main sources. First, mapping of the ocean floors during and after the war revealed an incredibly complex topography: Tremendous trenches in the sea floor rimmed the Pacific Ocean close to the coasts of Asia, Australia and North and South America. Long, sinuous midocean ridges encircled Antarctica and bisected the Atlantic and Indian oceans and the eastern part of the Pacific Ocean. Volcanic islands, such as Iceland, straddled these ocean ridges.

Second, the ocean ridges and trenches were hot, active zones, shuddering with earthquakes and erupting in volcanoes. The record of earthquake seismic waves confirmed what the early explorers had suspected: The earth's crust was not anchored. Instead, thirty to sixty miles below the surface, the crust slithered on a zone of hot, liquid rock.

Third, the age and magnetism of the ocean-floor rocks showed a strange pattern. For example, sediments on the floor of the Atlantic Ocean increased in geologic age outward from the mid-Atlantic ridge to the east coast of North America and to the west coast of Africa. The iron constituents of these sediments also preserved a magnetic blueprint from the past. Some of the sediments were "normal," that is, magnetized toward magnetic north, which is the position of the magnetic pole today. Other oceanic rocks were "reversed," or magnetized toward magnetic south, as if the magnetic pole had once sat near the South Pole. Apparently, the magnetic field (and magnetic poles) of the earth had flipflopped many times during the past. Most intriguing was the pattern of normal and reversed rocks along the ocean floors: They were aligned in a series of alternating stripes parallel to the midoceanic ridges. For example, from the mid-Atlantic ridge to the east coast of North America these parallel bands of rock registered a magnetic reading of normal . . . reversed . . . normal . . . reversed . . . normal . . . reversed . . . and so on when exposed to a magnetometer.

This evidence—matching fossils and coastlines, earthquakes, volcanoes, magnetism, ocean ridges and trenches—fit together in a revolutionary theory called plate tectonics. Its tenets overturned orthodox geology and its notions of a stable stodgy earth and recast the landmasses, oceans and interior of the earth as cauldrons of geologic chaos.

According to plate tectonics, the earth's crust is divided into seven major pieces ("plates") named for the major continents or oceans that ride on them: the North American, South American, Eurasian, African, Australian, Antarctic and Pacific plates. The boundaries between plates are the huge ridges and trenches on the ocean floor. The plates slide slowly, an inch or so a year, over the surface of the earth on a plastic layer of molten rock, propelled by the heat of convection currents from below the crust.

When plates meet, they collide, slide by, or one dives beneath another. Plate collisions elevate mountains. Plates sliding by one another cause earthquakes. A plate that dives under another—a process called subduction—is melted by great heat and pressure as it sinks down into the depths of the earth and causes massive earthquakes and volcanic eruptions on the surface above. Essentially, old crust dies in these fiery subduction trenches. New crust is born at midocean ridges, at the other edge of the plate, where molten lava wells up, spreads out on either side of the ridge, hardens and freezes the magnetic polarity in force at the time—normal or reversed.

The incessant theme of plate tectonics is recycling. Plates move as on a conveyor belt from midocean ridges—where new crust is added—to ocean trenches, where the crust is plunged into the earth's interior, heated, melted and eventually ejected as molten rock onto the surface. The net amount of crust on the earth doesn't change much, only the topography.

### Prediction 4: Violent earthquakes in the Americas and Asia

The Pacific plate is scraping slowly, inexorably, two to three inches a year along the western edge of the North American plate. The temporary, spasmodic suture between the two is the San Andreas fault. Amid constant rumbles and devastating earthquakes, the Pacific plate plods north and west, taking with it Los Angeles, the Mojave Desert and the coast ranges. In 10 million years, Los Angeles will be a suburb of San Francisco, and in 50 million years it will be situated off the west coast of Canada, where Hollywood will become nationalized as a public utility.

Farther south, the Pacific plate is plunging into a trench under the western edge of the South American plate. As a result, the Andes have been thrust upward over the past 80 million years by the folding and heating of the crust along the western edge of South America, and by powerful earthquakes and explosive volcanoes—violent processes that continue today.

Needless to say, these prophecies came to pass as predicted in 1983 (and each year since). On March 31, 1983, a powerful earthquake registering 5.5 on the Richter scale ripped through the Andes Mountains 350 miles south of Bogotá, Colombia. As reported in the news media, the 450-year-old city of Popayán collapsed, killing five hundred residents and destroying three thousand structures. Less severe earthquakes rumbled through parts of the Andes and Himalayas during February and March. Three shook southern California in April and May 1983 and one rattled a part of China earlier that year. Violent earthquakes rocked New Britain, an island in the South Pacific off the coast of New Guinea, in March 1983, and Japan in June. Both are near oceanic trenches (subduction zones) at the edge of the Pacific plate. Massive quakes hit Los Angeles and Alaska in 1987 and plate tectonics guarantees more earthquakes will rumble through these regions in 1988.

**Prediction 5: The Himalayas rise higher**

The Himalayas also owe their geological allegiance to moving plates. Around 70 million years ago, India wrenched away from Africa and drifted northward at a speed of about 3 inches per year. It doesn't sound very fast, but after 20 million years the Indian plate had drifted 60 million inches, or about 1,000 miles—the distance between Pittsburgh and Denver. India then rammed into Asia, causing the spectacular buckling and upheaval of the edge of the Asian landmass into the Himalayan Mountains. The Indian plate is still plowing into Asia, pushing the Himalayas higher an inch or so a year. Next year, conquerors of Mount Everest will have farther to climb.

**Prediction 6: Fiery volcanoes on Hawaii and Iceland**

Hawaii and Iceland were born in volcanism. Iceland straddles the mid-Atlantic ridge, where new crust wells up and is added to plates on either side of the ridge—the North American and Eurasian plates in the north and the South American and African plates in the south. Intense eruptions from these spreading ridges created volcanic islands, like Iceland 15 million years ago, or nearby Surtsey in 1963, both of which will continue to erupt. Iceland has almost two hundred active volcanoes.

Hawaii also regularly replays its fiery geologic history. Yearly eruptions of Kilauea eject millions of cubic yards of molten lava, adding new layers of volcanic rock to the Hawaiian terrain. The arc of islands is drifting slowly over a hot spot under the Pacific Ocean floor. Each island is merely the top of a towering oceanic mountain, built layer by layer over millions of years of volcanic outbursts until it pokes above the surface waters of the Pacific. Geologists in 1987 discovered a new volcanic island that will eventually emerge and be added to the Hawaiian chain.

**Prediction 7: North America goes west**

Back to Pangaea. The supercontinent was forged 300 million years ago when the world's landmasses drifted together and annealed. The geologic scars from those continental collisions still mark the earth's landscape: Ancient northern Asia smacked into archaic eastern Europe, elevating the Urals; Africa collided with North America, raising the Appalachian Mountains in the west and the Mauritanian range in the east. One hundred million years later (200 million years ago) the union of continents was broken. Pangaea splintered over hot, spreading ridges, first into two continents—Laurasia in the north (containing present-day North America, Greenland and Eurasia) and Gondwana in the south (South America, Africa, Australia and Antarctica). Then, 135 million years ago, North America wrenched free from Eurasia, and South America from Africa, along a ridge that now bisects the Atlantic Ocean floor. The Americas drifted westward, Eurasia and Africa drifted eastward, each on its respective plate, creating the Atlantic Ocean. Seventy million years ago, while dinosaurs were becoming extinct, India and Madagascar divorced Africa. About 20 million years later, Australia was severed from Antarctica and India whacked into Asia.

As North and South America drifted westward, the Atlantic Ocean kept expanding at the expense of the Pacific. This oceanic imperialism continues today because the pieces of the earth's crust haven't stopped moving. So, predicting a westward shimmy of North America is like predicting the Yankees will fire their manager during the course of the season. According to precise geophysical measurements, the North American plate moves an average of a centimeter a year. By the 1988 All-Star break, Billy Martin will be gone and North America will be halfway through its half-inch journey. At this rate, the Atlantic Ocean will be about 100 miles wider in 15 million years. I predict that transatlantic fares will go up.

# CHAPTER 24

XXI.—Ideal scene in the Lower Cretaceous Period, with Iguanodon and Megalosaurus.

Iguanodon *(left) and* Megalosaurus, *the first dinosaurs found, were reconstructed initially as four-legged. Later discoveries indicated that both were two-legged and that the nose horn on* Iguanodon *was really a bony thumb spike. (From Louis Figuier,* The World Before the Deluge, *D. Appleton and Co., New York, 1867)*

# Body Double: Duplicating Dinosaurs

At four o'clock in the afternoon of December 31, 1853, twenty-two distinguished British scientists sat down to a New Year's Eve dinner in south London. If the early dining hour was unconventional, so were the invitations to the feast. They came engraved on a fossil wing bone of an extinct pterodactyl and summoned the guests to the new Crystal Palace gardens where London's Great Exhibition of 1851, the first World's Fair, had been moved and reassembled. The scientists found themselves in a banquet hall as remarkable and bizarre as the bony invitations: They were to dine inside the yawning belly of an enormous beast, a colossal, lifelike, cement mock-up of the dinosaur *Iguanodon.*

With such flamboyance began our passion for duplicating dinosaurs. Fed by the awesome size of the beasts, the spectacle seems excusable, almost proper: The staging can't help but be grand nor the undertaking prodigious. It is the price of our romance with these bygone giants, an affair with epic monsters consummated from the safety of 65 million years.

Bringing dinosaurs to life also conjures a magical resurrection, a wonderworking that, as the White Queen told Alice, lets one believe as many as six impossible things before breakfast. Only an immense sleight of hand could be behind the cement Crystal Palace *Iguanodon;* or behind eighty-four-foot plaster skeletons of *Diplodocus carnegii* in London, Leningrad, Vienna, Berlin, Madrid, Paris, Bologna, La Plata, Munich and Mexico City; or a teenage *Camarasaurus* frozen in fiberglass near Jensen, Utah; or the robotic body doubles of *Apatosaurus, Tyrannosaurus, Triceratops* and *Stegosaurus* weaving and trumpeting through Pittsburgh.

The hosts for the New Year's Eve dinner in 1853 in London were Richard Owen, the renowned anatomist and paleontologist at the Royal College of Surgeons, and Waterhouse Hawkins, an eccentric sculptor and illustrator with an artistic fondness for the esthetics of the extinct. Hawkins had fashioned the concrete life-sized restoration of *Iguanodon* following Owen's anatomical blueprints, which were the only ones to follow in 1853. The same Owen had first coined the term "Dinosauria" in 1841 for the nine "terrible lizards" recognized at the time as vanished reptilian behemoths from ancient eras.

The banquet celebrated this *Iguanodon* incarnate, a monstrous metamorphosis in stone, cement, bricks, tiles and iron. Nowhere else had reconstructions of extinct beasts been attempted on such an imposing scale. Near the *Iguanodon* on the Crystal Palace grounds loomed a second concrete giant, the dinosaur *Megalosaurus,* amid other models of extinct reptiles. The two cement dinosaurs looked about as nimble as cows in a cage. The *Iguanodon,* girded by four ponderous legs, had huge, clawed feet, a skin of diamond-shaped green scales, and a horn on its nose—a rhinoceros in lizard's clothing. Only its gaping belly, now gorged with distinguished dinner guests, awaited the closing sutures of tiles, mortar and paint. Owen presided at the head of the table, which by apt design, was in the head of the dinosaur. Hawkins and the other guests lined the abdomen. As they toasted the imminent popularity of the restored *Iguanodon*, little did they suspect that their dinosaurian banquet hall would later become a monumental curiosity, an icon to anatomical blunder and fanciful reconstruction.

The Owen-Hawkins versions of *Iguanodon* and *Megalosaurus* were, as they say in the south, bodaciously wrong. This was not surprising, seeing that they had extrapolated the body, boots and britches of the dinosaurs from only the few scattered bones of the extinct reptiles known at the time. Complete skeletons found later testified that *Iguanodon* was not four-footed, but walked upright on its large hind limbs, and had short hands with a bony spike over each thumb, the same spike Owen had mistaken for a nose horn. *Megalosaurus* too turned out to be a two-footed dinosaur that Owen had plunked down on all fours. In 1936 the Crystal Palace burned down, but today these concrete mutants still rule the gardens, fabulous creations evolved to the fictional.

U.S. Highway 40, the main street in Vernal, Utah, runs past a stately red-brick building that looks like a bank but houses a small, tidy museum, the Utah Field House of Natural History. Outside the museum, three gray concrete dinosaurs, *Ceratosaurus*, *Camarasaurus* and *Stegosaurus*, stand at close quarters on a concrete pad behind a chain-link fence. It is a habitat far removed from their lush, tropical Jurassic digs, and an ecologic soiree far removed from the daily Jurassic adventure of eater versus eaten. Only through the dark prophetic glass of Isaiah would *Cerato-*

*saurus*, a voracious hunter, dwell with a juicy lamb like *Camarasaurus* or lie down with a tasty kid like *Stegosaurus.*

Rocks and time unite this dinosaur assembly. The skulls and skeletons of all three are preserved in the sands and muds of the Morrison Formation, the Jurassic rock layer below Vernal. The Morrison, laid down 135 million years ago in broad sheets across huge floodplains, covers a good deal of Utah, Colorado and Wyoming today. About fifteen miles east of Vernal on U.S. 40, near a wide spot in the road called Jensen, Utah, a huge chunk of the Morrison hurtles out of the ground at 60 degrees and begins its national acclaim as the rock of dinosaur ages.

Jensen is at the entrance to Dinosaur National Monument, a site where, on August 19, 1909, Earl Douglass, a Carnegie Museum paleontologist, discovered eight tailbones of a giant dinosaur coming out of the tilted Morrison rocks. The tailbones led to the hipbones and eventually to most of the skeleton. The Utah dinosaur, a new species, was named *Apatosaurus louisae* in honor of Louise Carnegie, Andrew's wife. The *Apatosaurus* excavations uncovered a second huge dinosaur, then a third, then more: Douglass had found a geologic graveyard, an ancient sandbar of a Jurassic river where dinosaur carcasses had washed in and been buried over hundreds of years.

One of those carcasses was a juvenile *Camarasaurus*, found by Douglass in 1919. If it had survived to full adulthood, the teenage dinosaur would have ballooned to a length of fifty-nine feet and a weight of twelve tons. As it is, this seventeen-foot immature *Camarasaurus*, Carnegie Museum Vertebrate fossil 11338, is the most perfect skeleton of a sauropod dinosaur ever found. Sauropods were those gargantuan vegetarian dinosaurs that mimicked bridges: Four huge, pillarlike limbs support a long, arching span of head, neck, body and tail—*Apatosaurus* (aka *Brontosaurus*), *Diplodocus*, *Brachiosaurus* and other behemoths.

The young *Camarasaurus* was exhumed in one solid slab of Morrison rock, shipped to the Carnegie Museum, exposed and put on display in Dinosaur Hall, its head and tail still flexed in classic death posture. Sixty-five years later, in 1985, Dinosaur National Monument requested a copy of this exquisite skeleton, a homecoming for its body double. The task of duplicating *Camarasaurus*, painstaking and monumental, fell to Norman Wuerthele, chief preparator in the Section of Vertebrate Fossils at The Carnegie Museum, and Chris Overand, a Pittsburgh artist with expertise in fiberglass casting.

A one-piece mold was impossible, given the size of the specimen. Instead, the dinosaur would be molded in seven sections and bolted together. Over the next year, Wuerthele and Overand stopped traffic in Dinosaur Hall as they made more and more of the juvenile *Camarasaurus* disappear behind mounting strata of Cementex #80, a rubber latex especially formulated for mold making:[1] ten coats of latex, a reinforcing layer of cheesecloth, ten more coats of latex, topped off with a covering shell of lightweight plaster reinforced with Nico Fiber, a fiberglass additive. Thirty gallons of latex encased the skeleton; each coat required two to three hours to dry. When the entire sandwich was peeled away from the *Camarasaurus*, the innermost layer of latex bore the mirror image of the ancient beast, the precise impression of bones, the faintest indication of scars, pits and grooves.

The dinosaur was poured and duplicated in polyester resin and fiberglass cloth in the old Duquesne Brewery on Pittsburgh's South Side, now home to a commercial casting company. Koppers Corporation donated the resin, PPG the rolls of fiberglass. The first *Camarasaurus* clone, frighteningly real, now lies among the other skeletons at Dinosaur National Monument. The second copy almost passes for genuine at the Los Angeles County Museum of Natural History. Fifteen thousand dollars buys the ultimate dinosaur collectible.

In the spring of 1899, Andrew Carnegie, smitten with the grandeur of dinosaurs, funded an expedition to the Morrison Formation rocks of Wyoming to find a dinosaur for his museum in Pittsburgh. A few months later, in a gully near Sheep Creek, Wyoming, Carnegie paleontologist Jacob Wortman and the museum's chief preparator Arthur Coggeshall found the first skeleton of what was later named *Diplodocus carnegii.*[2] The bones were removed from the rock, molded and cast at the Carnegie Museum in Pittsburgh, and in 1905, the first replica of *Diplodocus* was formally presented to King Edward VII at the British Museum in London. Over the next thirty years, *Diplodocus* casts were commissioned for museums in Paris, Berlin, Vienna, Leningrad, Bologna, Madrid, La Plata, Mexico City and finally Munich in 1934, where the plaster bones spent the war and the next thirty-two years still packed in their shipping crates in a basement storeroom.

In 1954, the original Carnegie Museum molds of *Diplodocus,* by this time tired and worn, were coaxed into begetting one more duplicate. The molds were sent to the Utah Field House in Vernal, where a cement replica of the skeleton was poured, assembled and mounted behind a fence on a cement pad, near Isaiah's peaceful flock of concrete dinosaurs from the Morrison Formation. The natural history museum in Las Vegas, Nevada, is gambling on molding the Vernal *Diplodocus* in the fall of 1988 for a copy of its own. The endless duplicating of *Diplodocus* seems to echo Andrew Carnegie's Shakespearean words in 1905 at the presentation of the first cast to the British museum: " 'Distribution shall undo excess' and both still have enough. We are to enrich each other . . . 'as one lamp lights another nor grows less.' "[3]

Forest Rogers, a Pittsburgh painter and sculptor, began duplicating dinosaurs at the Carnegie Museum about the time the *Camarasaurus* was getting its first coat of latex. She managed not to be seduced by the size of the beasts: Her replicas are in miniature, figures of giants whittled down to civilized, manageable proportions. Guided by the elegant dinosaur reconstructions in David Norman's *The Illustrated Encyclopedia of Dinosaurs*, Forest sculpted a Lilliputian Age of Reptiles, wax models imprinted with the intricate form and regal bearing of *Tyrannosaurus, Triceratops, Brachiosaurus, Diplodocus, Stegosaurus, Protoceratops, Parasaurolophus* and other dinosaurs. The journey from wax sculpture to plastic toy took the models from Pittsburgh to Hong Kong for molding, and to West Germany for casting and hand-painting.

Many commercial toy dinosaurs are dopey, stilted caricatures in dormant plastic. They are reproduced with the philosophy that our children's fervor for the beasts will forgive the lack of animation and anatomical accuracy. Until recently, the best plastic models on the market hailed from the British Museum of Natural History in London. But the English models are bested in liveliness and precision by Forest's plastic bestiary. Only three traits separate her bantam dinosaurs from Mesozoic reality: They are smaller, inorganic and bear a Carnegie Museum brand on their abdomens. About five dollars buys the poor man's ultimate plastic dinosaur collectible at the Natural History shop.

Another $7.95 buys ecological omniscience: a *Stick &*

*Learn Book* called *Dinosaurs and Other Creatures of Long Ago,* which Forest Rogers designed and illustrated with Kathy Borland.[4] The four "long ago" scenes paint an enchanted, whimsical Cretaceous world: a cheerful redwood forest inhabitated by *Iguanodon,* ankylosaurs, stegosaurs and a shrewlike mammal; a storybook swamp choked with ferns, palms, horsetails, cycads, turtles, dragonflies, snakes and birds; a parched, desert upland where *Tyrannosaurus,* two horned dinosaurs and a dome-headed dinosaur are fleeing for safety from a volcanic eruption; and the Kansas Seaway, an inland ocean devoid of dinosaurs but teeming with ammonites, fish, aquatic reptiles, flying reptiles, birds.

Inserted between these scenes are plastic pages littered with forty-five peel-off, stick-on illustrations of Cretaceous creatures: eighteen different dinosaurs, ammonites, twelve-foot bony fish, long- and short-necked plesiosaurs, mosasaurs, porpoiselike ichthyosaurs, giant sea turtles, pterosaurs, toothed birds and mammals. The reader now becomes a paleontological czar, master of the Cretaceous terrain, able to marry animals to plants, populate landscapes and create ancient ecosystems by whim, fiat, peel and stick.

In the world of facsimile dinosaurs, few restorations are more dramatic than those of Dinamation, a company that manufactures and exhibits a pack of prehistoric behemoths that twist, swivel, roar and honk. They are wildly popular: Three million people saw Dinamation's versions of *Apatosaurus, Tyrannosaurus* and kin in 1986, 4.5 million in 1987. Four groups of Dinamations's dinosaurs and other archaic creatures crisscross the country, hopping from museum to museum, a traveling road show of extinct marvels. This year, eleven reptilian marvels from the Dinamation stables descended on the Carnegie Museum for a summer run under the marquee DINOSAURS ALIVE!

If spectacle and sheer size sold the Crystal Palace dinosaurs, the hook behind the Dinamated ones is just that: kinetics and sound. The textured, painted rubber skin of these robotic dinosaurs covers steel plates, hoses, rods, pistons, gears, bushings and wires—synthetic organs driven by the hiss and whoosh of compressed air. *Tyrannosaurus* roars, raises its menacing head and stiff tail and swings its body around to survey the crowds. *Stegosaurus*, sniffing left and right, paws its right front foot. *Apatosaurus* yowls, its sinuous neck arching from side to side. *Corythosaurus* honks like a Buick,[5] rears up on its hind legs for a better view and sees a bellowing *Triceratops* slowly lower its huge head shield and horns for a bull-like charge.

Dinamation's creatures have also evolved. The first generation that toured the country were cute, almost cuddly—the novelty of motion excused the cartoon countenance. The newest among the current crop abandon user-friendliness for a fierce realism. *Tyrannosaurus*, formerly a brown, stiff, ill-proportioned tripod resting on legs and tail, is now a lean, mean, reptilian machine, its body pivoting over tensed, muscular legs ready to attack. But, as in the real world, some of the intended victims, like Dinamation's *Corythosaurus*, have not quite kept evolutionary pace.

Along U.S. Highway 59 in East Texas rise the billboards for Bean's Dinosaur Gardens. A few years ago, Mr. Donald Bean, a retired carpenter, fulfilled his lifelong dinosaur passion by sinking $100,000 and eleven life-size reptilian replicas into the woods outside Moscow, Texas, a small town between Corrigan and Seven Oaks, north of Houston.[6] In contrast to the systematic formality of Owen's Crystal Palace, the Dinosaur Gardens is a folksy roadside park. Also folksy is its dinosaurian taxonomy: It lumps dinosaurs, such as *Triceratops,* with other extinct nondinosaurian denizens such as the fin-backed reptile *Dimetrodon,* and the long-necked sea monster *Elasmosaurus,* the same plesiosaur that tourists to Scotland like to think they see in Loch Ness. Mr. Bean, like Owen and Hawkins, was bewitched by size. The duplicate *Triceratops,* built from fiberglass, chicken wire, urethane and plaster, weighs three thousand pounds. The *Elasmosaurus* is forty feet long.

None of this, however, seems impressive enough to make many east Texans who travel U.S. 59 stop at Moscow. Mr. Bean's dinosaurs are in economic danger of extinction, honing the moral saw that art will imitate life.

In the end, the same saw wounded Waterhouse Hawkins. Americans first discovered Hawkins in 1866 when the Ward's catalog for that year offered a petite set of his Crystal Palace creatures. Two years later, more than a decade after the *Iguanodon* dinner in London, Hawkins visited the United States and was invited to New York by the "Commissioners of the Park" to entertain a stunning proposal: re-create a bestiary of North America's extinct giants right there in Manhattan in Central Park.[7] Hawkins accepted.

In deference to cement physiology and New York winters, the gigantic models would not be out in the elements but installed in a museum in the park. Although it was to be called the "Palaeozoic Museum," its inhabitants would be vanished giants from the Mesozoic and Pleistocene: two duck-billed dinosaurs (a twenty-six foot and thirty-nine foot *Hadrosaurus*); two carnivorous dinosaurs (*Dryptosaurus*); a plesiosaur (*Elasmosaurus*); two huge armadillos that resembled armored dirigibles (*Glyptodon*); two enormous ground sloths (*Megatherium, Megalonyx*); a mastodon; a mammoth; and a carnivorous mammal, perhaps *Smilodon*, the sabertooth cat.

The museum's foundation was excavated and built at a cost of $30,000 in Central Park near what is now Central Park West and Sixty-third Street. Plans for the museum building called for a cast-iron and glass affair with an enormous arching roof and a price tag of $300,000. Hawkins, in a temporary studio on the second floor of the Armory, began fashioning his restorations.

But in 1870 the political climate cooled for these archaic creatures. Tammany Hall and William "Boss" Tweed's corrupt "Tweed Ring" regained control of New York's parks and much of the city and lowered the boom on the Palaeozoic Museum. Two of Tweed's hatchet men, Peter Sweeney and Judge Henry Hilton, camouflaged in moral righteousness, spread the propaganda that the antediluvian monsters would foster antireligious attitudes. Pious hoodlums incited by Hilton broke into and trashed Hawkins's studio.

The museum perished. Hawkins's gigantic molds and counterfeit creatures were dumped into the open foundation and bulldozed over. The burial is thick with paleontological irony, imparting to the dinosaur body doubles a fate shared only with their extinct organic forebears. These skeletons of models of skeletons that lie preserved in the strata beneath Central Park are now proper specimens themselves, archaeological artifacts waiting for a different museum.

# CHAPTER 25

*A scientifically literate view of the moose in 1792. (From Thomas Pennant,* Arctic Zoology, *Robert Faulder, London, 1792)*

# Scientific Literacy

Literate societies like to produce lists of knowledge. Some lists are long: dictionaries, atlases, encyclopedias. Some are short: heroes of France, Albanian resorts, Canadian statesmen. Some lists confuse a private snobbery for the public interest but are merely good evidence that any fool can produce one: what's "in," what's "out," what's trendy, what isn't.

A new list of knowledge surfaced recently. It's called *Cultural Literacy: What Every American Needs to Know*, by E. D. Hirsch, Jr.,[1] a professor of English at the University of Virginia. Part I of his book is an indictment of American educational curricula. Apparently, most of us don't know an *A* from a bull's foot. For example, these blunders were committed by American students, from eighth-graders to college juniors, when quizzed on their cultural savvy:

- Toronto is a city in Italy.
- The Alamo is a Greek epic written by Homer.
- Christ was born in the sixteenth century.
- Socrates was a great American Indian chief.

- Latin is the language spoken in Latin America.
- The Great Gatsby was a famous musician.

In another test, two thirds of seventeen-year-olds didn't know that the Civil War took place between 1850 and 1900, and half didn't recognize Winston Churchill or Joseph Stalin.[2] Such is the evidence of widespread cultural illiteracy among America's youth, according to Hirsch.

Part II of his book is part of the cure—a list of 4,500 terms, dates, concepts and ideas, sixty pages long, that every American ought to have in his/her knowledge box, ranging from "abbreviation" and "abominable snowman" to "zoning" and "Zurich."

Well, calling the Alamo a Greek epic is a bit of a clanger all right. But, details aside, the saga of the Alamo is basically like the *Iliad*, just substitute Davy Crockett, Sam Houston and Santa Ana for Hector, Paris and Achilles. It's also in a language we can understand. Speaking of language, if Latin isn't spoken in Latin America, wherefore Latin America? Why not Spanish America? And, concerning Latin Toronto, most Canadians will tell you that they'd gladly give Toronto to Italy, considering what passes for hockey at Maple Leaf Gardens.

Kidding aside, what's the use of knowing all 4,500 of Hirsch's cultural basics, when cultural literacy isn't necessarily a passport to reason. Consider the 1987 ruling of the U.S. Supreme Court striking down the constitutionality of teaching "scientific creationism" in Louisiana's public schools, in which two nays were registered by Justices Scalia and Rehnquist.[3] Why should we expect a seventeen-year-old student to recognize Stalin, when two Supreme Court justices can't recognize Darwin?

Perhaps Rehnquist and Scalia will read *Cultural Liter-*

*acy* and its 4,500 terms and names, including Charles Darwin. The sixty-page list doesn't pretend to be the *Encyclopedia Britannica.* One reviewer of *Cultural Literacy* asked, why "Zurich" and why not "Zimbabwe"? Well, for that matter, why "Zimbabwe" and why not "zoology"? But, as the reviewer admitted, quibbles over this or that term miss the point. There is a larger role for this lineup of cultural kernels: to prescribe that minimum curriculum of facts and concepts fundamental to our society—a cultural meat and potatoes.

Nevertheless, I'd like to add a few quaint, lesser-known natural-history terms to Hirsch's recipe for our lowest cultural common denominator. Perhaps someone will also send *Scientific Literacy* along to the justices.

**aa:** In addition to being the last two letters of countless automobile associations and the acronym for Alcoholics Anonymous, aa is a Hawaiian term for a lava with a rough, clinkery, jagged surface. The Hawaiian language is rich in words describing types of lava, courtesy of having developed on an island arc forged by undersea volcanoes.

**anemochore** ("a-knee-mow-core"): The scientifically literate way of saying "dispersed by wind" to describe, for example, the pernicious behavior of dandelion seeds. Similarly, *hydrochore* is biospeak for "dispersed by water." Both terms are rooted in Greek, which too often is a convenient excuse for unneeded jargon.

We can't help it if some naturalists persist in manufacturing convoluted names based on classicized ideas (eg., anemohydrochore—ergo, wind *and* water dispersion) or unpronounceable places (see MAASTRICHTIAN). These sorts of scientific terms alienate an already suspecting public. And what do they suspect? That the scientific world has a moose in charge of its PR department.

If it were up to me, all great ideas would either be reducible to a two- or three-letter/symbol designation—as in AD, BC, AC, DC, MPH—or be consigned to the monkey-at-the-typewriter heap of life. EVolution would be EV; GRavity, GR; ElectroMagnetism, EM; Quantum Mechanics, QM; Art, ???; Cubism, ▱ ; EXpressionism, EX; Abstract Expressionism, @&; Pointilism, . . . ; Minimalism, 00; Punk Rock, PR; Continental Drift (and Compact Disc), CD—you get the idea. When there is overlap—PR and PR, CD and CD—then it's merely a replay of the homonym problem in literature and the species problem in paleontology: The context will reveal the meaning.

**apomorphy** ("ape-oh-more-fee"): This is another fine example of the PR moose. Apomorphy has nothing to do with apes, although apes have apomorphies. An apomorphy is any evolutionary novelty, any advanced physical trait possessed by an organism. For example, the large brain of *Homo sapiens* is an apomorphy with respect to the smaller brain in *Australopithecus afarensis,* the extinct hominid species to which "Lucy" belongs. Two or more organisms with the same apomorphy are said to share a synapomorphy, which sounds like an incurable disease, but is the doppelgänger way of saying that these organisms share the same evolutionary novelty inherited from the same ancestor. To continue the example, *Homo sapiens, Homo habilis* and *Homo erectus* all have larger brains compared to *Australopithecus afarensis* and this shared advanced trait is one of many synapomorphies testifying to their evolutionary kinship.

This user-friendly lingo was coined by a German naturalist, Willi Hennig, to accompany his precise, ironclad system of deciphering and expressing evolutionary relationships among organisms, called cladistics. The flip side of apomorphy is plesiomorphy ("plea-zee-oh-mor-fee"), meaning "primitive trait," and its twin, symplesiomorphy ("sharing a primitive trait").

One of the best examples of a symplesiomorphy is the need shared by German naturalists to invent a sausage-belt vocabulary where simple terms already exist. One of those simple terms is "unique trait," for which Hennig invented "autapomorphy" ("ought-ape-oh-more-fee"). An autapomorphy is an apomorphy that only one species developed and does not share with any other species, a one-of-a-kind feature. Man has many autapomorphies, among them the capacity for mistaking terminological gobbledegook for scientific literacy.

**badlands:** A rugged, rocky, barren landscape of weathered canyons and mesas, virtually devoid of vegetation. In the American and Canadian West, badlands are bad for ranchers, cows and sheep. But to paleontologists, badlands are the fertile terrain of past life, continuously eroding out the petrified remains of the animals and plants from an ancient earth.

**behead:** A geological term for the cut-off and capture of the upper part of a stream by another stream. This is the definition Marie Antoinette would have preferred.

**blackjack:** 1. a poor, thin bed of coal; 2. a crude black oil used to lubricate the wheels of mine cars; 3. a crude card game used to lubricate the bank accounts of casino owners.

**bort:** A diamond too flawed or badly colored for jewelry. Buying your fiancée a bort may abort your engagement.

**bullion:** On Wall Street, the traditional hedge against inflation; in geology, a nodule in coal formed around a fossil plant.

**cactolith:** A stringy type of lava flow. This is a simple enough definition, but the classification of lava flows is apparently far from simple. Here is the definition of cactolith in a college geology textbook: "A quasi-horizontal chonolith composed of anastomosing ductoliths, whose distal ends curl like a harpolith, thin like a sphenolith, or bulge discordantly like an akmolith or ethmolith."[4] Is this literacy of any kind, cultural or scientific?

**carat:** A French unit of weight for diamonds, equaling 200 milligrams. A carat is also a measure of the fineness of a gold alloy, with the ultimate being 24-carat gold or fine gold. This marriage-by-carat of gold to diamonds is one of convenience to the jeweler and financial consternation to the groom. Fourteen-carat gold may be only fourteen-twenty-fourths gold, and a one-eighth-carat diamond may be microscopic, but together they are carrot enough to attract betrothal and debt.

**Le Chatelier's principle:** A shift in the conditions of a system causes a shift in equilibrium that eventually shifts the system back to its original conditions. This is why France hasn't progressed very far.

**Chernozem** ("chair-no-zem"): Russian for "black earth." Chernozem, a soil-science term, describes a black soil in the Soviet Union rich in organic debris (humus), and is not to be confused with Chernobyl, a black soil in the Soviet Union rich in nuclear debris.

**claibornian:** A formal period of earth history during the middle of the Eocene Epoch, about 50 million years ago, when, it is suspected, Craig Claiborne first discovered the eggbeater. It was also a geologic age when lush forests grew in the Arctic, when primitive horses, rhinoceroses and primates roamed Wyoming, and when India was an island continent cruising on its continental plate toward a spectacular collision with Asia.

**coprolite:** Not to be mistaken for a brand of "lite" beer from Copro Breweries, a coprolite is any fossil dung.

**Coriolis effect:** The effect generated by the earth's rotation on its axis, which is blamed for many things, including causing water in a bathtub to drain with a right-handed spiral in the Northern Hemisphere and a left-handed spiral in the Southern Hemisphere. It is named for Gaspard G. de Coriolis, a French mathematician who lived during the French Revolution, from 1792–1843, when bathtubs were in short supply (except in the Marat household), and when bathtubs with drains were an object of desire.

The Coriolis effect results from two forces: (1) The speed of the earth's rotation is higher at the equator than the poles and thus deflects any northbound bodies (ocean currents, missiles, wind systems) eastward and southbound ones westward (in the Northern

Hemisphere); (2) the earth's rotation (from west to east) generates a centrifugal force that deflects eastbound bodies southward and westbound ones northward.

**cow's bezoar:** A hard mass found in the stomach or intestines of some animals, usually ruminants (cud chewers), and once thought to be an antidote for poison. I first learned of cow's bezoar from Dave Love, whose formal title is geologist with the U.S. Geological Survey in Laramie, Wyoming but who is known to his colleagues as the "grand old man of Rocky Mountain geology."[5]

During a recent trip to the People's Republic of China, Love came down with Chiang Kai-shek's revenge. At a rural clinic he was given a bottle of gigantic pills and told to take one every few hours. He did so for a day or two, with no effect. Back in the United States, Love had a friend translate the ingredients listed on the bottle in Chinese: powder of rhino horn and cow's bezoar.

**Cuvier's principle:** Different features of organisms, such as teeth and feet in mammals, often evolve in step with one another and seem associated. This general observation predates Cuvier—"foot-in-mouth" disease, for example, has afflicted public figures ever since speech overtook thought. Cuvier's principle is well documented in the evolution of many organisms, including fossil horses, beginning about 55 million years ago with a collie-sized, four-toed "dawn horse" called *Hyracotherium* and ending in *Equus*, the one-toed living horse. Between these two is a story of changing teeth and feet. The teeth in various horse lineages became progressively modified into higher, larger, more complex dental batteries for more efficient processing of vegetation; concurrently, the feet became longer and emphasized one toe (the third) at the expense of the others, which became reduced and lost—all this for faster escape from predators and more efficient standing while feeding.

**exfoliation:** A two-page advertisement in a recent *New York Times Sunday Magazine* heralded a new line of nine skin-care products. The basic elixir is a mud concoction from northern Italy, which, the ad copy proclaims, has been rejuvenating weathered Roman skins since 400 B.C. Eight of the nine treatments promise to cleanse, restore, activate, energize, revitalize and hydrate the aging epidermis. The ninth promise steps over the boundary between natural hype and natural history: The mud "gently exfoliates the body, leaving the skin surprisingly tingly."

Well, real exfoliation is seldom gentle. The only surprise in store for the exfoliatee is that it would leave the body with no skin at all, tingly or otherwise.

Exfoliation is one of the natural processes of weathering. Its literal meaning is to strip of leaves. But, also through exfoliation, trees are stripped of bark, and rocks lose their outermost layers in concentric sheets. Exfoliation is especially common in moist climates. Water, with tiny amounts of dissolved carbon dioxide, seeps into cracks and crevices in rocks and causes expansion in the outside layers. These layers eventually separate and peel off in curved hunks. Years of exfoliation denude trees and transform angular boulders into smaller, rounded ones.

Exfoliation, and all other forms of weathering and erosion, are the price paid by the earth's terrain when it is not in complete equilibrium with its environment.

I imagine that the price paid for the Italian mud will only exfoliate your wallet.

**fumarole:** A hole or vent through which fumes or vapors issue, common in the earth's crust and in politicians up for reelection.

**geophone:** A wonderful name for a seismometer. Geophones listen for and record ground motion, usually caused by earthquakes.

**harmonica:** A bonehunter's nemesis. It describes a fossil mammal jaw found without any of the teeth preserved. The jawbone with the row of empty root sockets resembles a harmonica.

**horse latitudes:** Two belts of high air pressure located between 25 and 30 degrees on either side of the equator, providing calms, light winds and clear weather. The trade winds blow from the horse latitudes, which, accordingly, would have been named "bull latitudes" or "bear latitudes" had the stock-exchange people been in charge. As it is, the horse latitudes were christened by early Spanish sailors who were transporting horses to America and the West Indies. The calm seas of the horse latitudes slowed the sails and severely prolonged the voyage. During the ensuing water shortage, the first to be tossed overboard were the horses.

**Hooke's law:** Stress is proportional to strain, a simple observation, which, one would think, would make much of psychoanalysis obsolete.

**ideal gas:** The kind of gas that used to cost 25 cents a gallon. In physics, ideal gas is "perfect gas," i.e., it fully obeys the gas laws, especially Boyle's law ("the pressure of a gas is inversely proportional to its volume").

**jökull** ("yokel"): An Icelandic term for a glacier consisting of a large ice sheet.

**jökulhlaups** ("yo-kah-haul-ops"): A flood resulting from the breakup of a jökull dam.

**juvenile water:** Water from the interior of the earth that has previously never been in the atmosphere (as vapor) or on the surface (as liquid). When it reaches the surface, juvenile water gets bottled as a trendy French brand of designer water. What is difficult to swallow is the ad copy for Perrier, which is less hyperbole than an appeal to Stone Age chic. It is marketed as "Earth's first soft drink." Furthermore, the ads intone, "Prehistoric cuisine was salt free . . . because stone-age man didn't know there was such a thing as salt. Or high blood pressure either." And the punch line: "His [Stone Age man's] favorite libation came from a pure, sparkling, naturally salt-free spring."

The Stone Age, the earliest period in man's cultural evolution, lasted from about 2 million to 5,000 years ago; during it *Homo erectus* and various groups of early *Homo sapiens* (Neanderthals, Cro-Magnon and others) fashioned an assortment of stone tools for use in their hunting and gathering existence. If Stone Age peoples didn't know about high blood pressure, it was because they never lived long enough to incur it. Judging from preserved skeletons, their average life expectancy was probably not much above thirty-five or forty. Many northern Neanderthals, especially children, lived only long enough to die of rickets, caused by a deficiency of vitamin D in their diet.

Knowledge of salt is another matter. Stone Age people, like New Age perspirers, must have tasted their own salty sweat and discovered the spice in their skin. As hunter-gatherers, they subsisted on wild game, vegetation and fruit, a good deal of which is not salt-free. Ethnographic studies of modern aboriginal hunter-gatherers reveal that the internal organs of game animals, including the salt-rich kidneys, are prized morsels. Mammoth, reindeer, buffalo and rabbit kidneys wre probably high on the Neanderthal list of gourmet delicacies. And the blood of their prey, which is rich in dissolved sodium, may have been their second favorite libation.

Perrier's Stone Age spring may be "pure and primeval as anything," but, ironically, most natural springs are usually much richer in dissolved salts than streams and rivers. They are fed by underground water, which flows and percolates through mineral-rich sediments before reaching the surface.

**lab animal:** An animal that gets injected with a drug and produces a scientific paper. Rats and mice are the most prolific lab animals.

**lagomorph** ("lag-oh-morf"): The order of mammals containing the rabbits, hares and pikas, and a term bearing no relation to the various "morphs" in Hennig's cladistic vocabulary (see APOMORPHY). Rabbits can end life in stews or pregnancy tests or worse: It usually takes dozens of rabbits to make one mink coat.

The rabbit, of course, has been borrowed by one corporation for its Rabbit. The subtle message to the consumer is that he is buying an organism, a pet, not a machine. The same principle gave us Impalas, Jaguars, Cougars and Lynx as well as the fruits of technology: Apple computers, Peachtree software, and Lemon power-surge suppressors. Elk sell insurance for Hartford, tigers pump gas for Exxon, and bulls invest for Merrill Lynch. Except for the koala—a marsupial that flies Qantas—placental chauvinism is obvious in the world of organic commercialism. And, except for the Beetle, Turtle wax, Caterpillar's tractor, and Shell Oil's shell, mammals seem to have the lion's share of product imagery.

Perhaps there ought to be a commercial ethic whereby each class of animals is reserved for a class of commercial products. By tradition, cars should go to mammals—the Sloth and Platypus are still unclaimed—and computers should stick with plums, pears and nectarines. Airlines ought go to long-flying birds: Albatross Air; Tern Air. And oil companies should pay homage to the plants that gave their lives hundreds of millions of years ago for today's petroleum: Cycad Oil; Ginkgo Petroleum.

**Maastrichtian:** The last geologic stage of the Cretaceous Period, defined by strata in the Maastricht region of Holland. The end of the Maastrichtian also marks the end of the Mesozoic Era, the end of the Age of Reptiles, the end of the dinosaurs, and the end of innocence for meteorites (see METEORITE).

**meteorite:** Any natural object that falls to the earth from space. On November 18, 1492, thirty-seven days after Columbus fell into the New World, the oldest witnessed meteorite fall occurred at Ensisheim, France. The meteorite sighting was dismissed in France and the rest of scientific world—no one then believed that rocks could fall from the sky. They started believing three hundred years later, after meteorites had clonked down in France (Luce, September 13, 1768; Barbotan, July 24, 1790), Italy (Siena, June 16, 1794) and England (Wold Cottage, December 13, 1795). The stones that broke the camel's back were in a meteorite shower that bombarded L'Aigle, France, in April 1803. An investigative commission of the Royal Academy of Science in Paris finally concluded that stones do fall from the sky.

Meteorites come in three varieties: stony, iron and intermediate, graded according to their iron/nickel/silica content. Their fate too is varied. Small ones burn up in the earth's atmosphere. Large ones survive the fiery entry to pockmark the terrain with craters. Some of these large meteorites become sacred objects. The Black Stone of Mecca is a holy symbol to Muslims, as was the Casa Grande meteorite to Montezuma Indians. This 3,400-pound piece of space was found wrapped in coarse linen, a sign of reverence, in a Montezuma Indian temple.

One gigantic meteorite is accused of having killed off the last of the dinosaurs 65 million years ago, when it allegedly hit the earth, exploded and sent up a dust cloud that encircled the planet, blocked the sun's rays and created the equivalent of a nuclear winter. That the meteorite hit is possible. That it contributed to the end of the dinosaurs is improbable.[6]

**orogeny:** Orogenous zones should not be confused with erogenous zones, although both can experience uplift. Orogeny means geologic uplift, such as mountain building. The rise of the Rocky Mountains about 70 million years ago is referred to as the Laramide orogeny.

**photonasty:** In botany, photonasty describes those plants, such as the elegant evening primrose, that open and close in response to diffuse, nondirectional light. In life, photonasty describes the Hapsburg monarchs, punk rockers, sumo wrestlers and other autapomorphic humans (see APOMORPHY).

**primarrumpf:** In geology, an outward expanding dome of rock whose slow rise is matched by slow erosion. Among humans, primarrumpf is prima facie evidence of midde-age rumpf.

**slump:** The history of the Montreal Expos since their origin in 1969. In geology, a slump is the sudden slide of a massive chunk of earth, such as a hillside. When soil and rock tumble down a slope in bits and pieces, the process is called "creep" if it occurs slowly and "mass wasting" if it is faster and on a larger scale.

**species** (plus **lumper, splitter** and **splumper**): Put a gray-flannel suit on a Neanderthal, give him an attaché case, put him in a New York subway, and a lumper will call him *Homo sapiens*. Lumpers are classifiers who tend to downplay minor differences between individual organisms when they group them into species. As a result, lumpers classify the living and fossil world into fewer, broader species.

The opposite of lumpers are splitters, a group who claim to be more discriminating—to them even minor morphological variations among individuals warrant the recognition of discrete species. One look at our Neanderthal executive on the way to Wall Street—with the somewhat larger eyebrow ridges, the slight bulge at the back of the skull—and a splitter would change the initials on the attache case from

*"H. s."* to *"H. n."* for *Homo neanderthalensis.*

Another good example of the splitter-lumper dichotomy is bears. Lumpers group living North American bears into three species—*Ursus americanus* for the black bear, *Ursus arctos* for the brown or grizzly bear and *Ursus maritimus* for the polar bear. Splitters have divided the same bear world into twenty-six species in three genera *(Ursus, Euarctos* and *Thalarctos).*

Biologists run the gamut from splitters to lumpers, but most are middle-of-the-road "splumpers." All three are trying to answer the same question: What is the nature and extent of organismal diversity? What are the living species of animals and plants? Paleontologists have the same query about an ancient world: How many fossil species are indicated by the bones, shells, leaves and other petrified remains in the rocks? The problem is not in defining what a species is, but in applying the definition consistently.

Species, as classically understood, are groups of organisms that cannot interbreed with other groups; members of a species can reproduce (and exchange genes) only among themselves. The scientifically literate term for this situation is "reproductive isolation." This is a fine definition in theory, but wholly impractical. Go observe reproductive isolation in Amazonian beetles, or Rocky Mountain trout, or reef corals or fossil man. We will never know whether Neanderthals and modern humans could have bred and produced viable offspring.

But we can make intelligent inferences about whether they could have reproduced, or whether two beetles belong to the same species. The measure of that inference is morphology, the physical characteristics of animals and plants. If species cannot interbreed and exchange genes, they should differ consistently in some morphological trait and/or behavioral repertoire. In other words, morphologic gaps between groups of individuals imply their reproductive isolation and species status.

Trouble is, how much morphologic difference does a different species make?—a question that caused classifiers to speciate into splitters and lumpers. About a gnat's worth of difference, say splitters; a mammoth difference, say lumpers. Splumpers, who by nature put practicality before ideology, say that species should be recognized on readily and invariably distinguishing features.

Confounding this solution are the inevitable extreme cases. Splitters like to point to sibling species, which are reproductively isolated, as good species should be, yet are virtually identical in outward appearance and internal anatomy, differing only in morphological minutiae or ecological habits. They are common among insects (e.g., six sibling species of European malaria mosquitos) and fish and molluscs, but rarer in other groups. The converse of sibling species is a lumper's delight—those species that exhibit a prodigious amount of morphological variation between the sexes and across their geographic range. Good examples here are birds (Linnaeus originally described male and female mallard ducks as different species) and freshwater clams and snails (250 "species" of a clam called *Anodonta* turned out to be local variants of a single species), as well as certain mice, grasshoppers and many other animals.

The upshot is the age-old tension of theory versus practice. In theory, species are composed of genetic kin. In practice, species are groups of a morphologic kin. The bridge betwen the two—for splitters, lumpers and splumpers—is subjective, which makes taxonomy, the ordering of nature, both a science and an art.

**stratification:** The layered structure of rocks deposited by wind and water. Humans too are stratified, courtesy of their social systems, although this is hotly denied by the upper strata in Soviet society.

Fundamental to stratification is the Law of Superposition, which ardent creationists regard with aroused suspicion on two counts: It sounds more erotigraphic than stratigraphic; and, geologically speaking, it proves that the earth is much older than the 4,004 to 10,000 years they allow for the biblical "creation week." Nicolaus Steno, a Danish naturalist, formulated the Law of Superposition in 1669. It states the simple observation that layers of rocks are deposited through time one on top of the other, so that in a local sequence of rock layers, the lower strata are older, and the upper, overlying strata younger. Mine shafts are "stratavators"—the rocks are progressively older as one proceeds down the shaft.

Thick rock sequences, like ones exposed in the Grand Canyon, Arizona, and the Wind River Canyon, Wyoming, preserve hundreds of millions of years of earth history from bottom to top. The evolution of life is a part of that history, and fundamental to understanding it is the Law of Faunal Succession. It states that the species composing floras (plant communities) and faunas (animal communities) change steadily through time, so that rocks with the same fossil animals or plants can be considered to be of roughly the same age. This was deduced in 1799 by William Smith, who was termed the "Father of British geology" or, more colloquially, "Strata Smith." The reasons for change and turnover in the species of animals and plants through time was explained sixty years later by Charles Darwin's theory of organic evolution.

**transgression:** In religion, a sin; in geology, the invasion and covering of land by a shallow sea, usually caused when sea level rises or land subsides. Transgression is followed by repentance in religion, and by regression, the retreat of the sea, in geology. In the fable of Noah's Flood, one transgression brought on the other.

**Uintatherium** ("beast of the Uinta Mountains"): An extinct, gigantic, rhinoceroslike, hoofed mammal that roamed North America and Asia during the middle and late Eocene, about 45 to 50 million years ago. *Uintatherium* and its kin were not only the largest land animals of their time, but also the most grotesque. They had elephantine limbs and feet and huge, curving, scimitarlike upper canines (fang teeth) that were protected by bony flanges projecting down from the lower jaw. Their skulls were fantastic saddle-shaped affairs, with pairs of bony knobs protruding above the nose, eyes and back of the head. The ones over the nose may have supported horns (illustration, Chapter 13).

*Uintatherium* is so unlike any living beast that it became the unfortunate subject of one of the most fanciful and comical reconstructions of a fossil mammal. In 1886, in a popular book called *Life History of Our Planet,* William D. Gunning described *Uintatherium* as a hybrid monster, having the limbs of an elephant, the skull, molars and cheekbones of a rhinoceros, the front upper jawbone of a cud chewer, the canine and jaw motion of a carnivore and the minuscule brain of an opossum. The accompanying reconstruction confirmed the existence of yet another animal designed by a camel.[7] Gunning's *Uintatherium* boasted an elephantlike body with radar-dish ears, a short trunk, two canine tusks, four sharp horns above the nose and eyes, and velvet-covered mooselike antlers above the back of the skull (illustration, Chapter 12).

**zebra:** A donkey behind bars. The three species of zebra in Africa earn their recognition from their stripes: Burchell's zebra has a few wide black stripes; Grévy's has numerous narrow black stripes and a white patch on the rump; the mountain zebra resembles Grévy's but has transverse black stripes across the rump.

**zoology:** Hirsch should have reprinted the classic definition in Ambrose Bierce's *The Devil's Dictionary:* "The science and history of the animal kingdom, including its king, the House Fly *(Musca maledicta)*. The father of Zoology was Aristotle . . . but the name of its mother has not come down to us. Two of the science's most illustrious expounders were Buffon and Oliver Goldsmith, from both of whom we learn . . . that the domestic cow sheds its horns every two years."

# Notes

## CHAPTER 1: Dinosaur Plots

1. Robert T. Bakker, *The Dinosaur Heresies* (New York: William Morrow, 1986).

2. See Chapter 8: The *Archaeopteryx* "Hoax."

3. P. Dodson, Review of *The Dinosaur Heresies* by Robert T. Bakker, *American Scientist*, 75, September/October 1987.

4. A. Charig, *A New Look at the Dinosaurs* (New York: Mayflower Books, 1979).

5. *Science Digest*, June 1982.

6. L. W. Alvarez, W. Alvarez, F. Asaro, and H. V. Michel, "Extraterrestrial Cause for the Cretaceous-Tertiary Extinction," *Science*, 208, 1980, pp. 1,095–1,108.

7. L. Van Valen, "Catastrophes, Expectations, and the Evidence," *Paleobiology*, 10(1), 1984, pp. 121–137: "The Case Against Impact Extinctions," *Nature*, 311, 1984, pp. 17–18.

8. R. E. Sloan et. al., "Gradual Dinosaur Extinction and Simultaneous Ungulate Radiation in the Hell Creek Formation," *Science*, 232, 1986, pp. 629–633.

9. A. Hallam, "End Cretaceous Mass Extinction Event: Argument for Terrestrial Causation," *Science*, 238, 1987, pp. 1237–1242.

10. W. H. Zoller, et al., "Iridium Enrichment in Airborne Particles from Kilauea Volcano: January 1983," *Science*, 222, 1983, pp. 1118–1121.

11. C. B. Officer and C. L. Drake, "Terminal Cretaceous Environmental Events," *Science*, 227, 1985, pp. 1161–1167.

12. *Newsweek*, July 13, 1987.

## CHAPTER 2: The Naming of the Shrew

1. Ernst Mayr. *The Growth of Biological Thought: Diversity, Evolution and Inheritance* (Cambridge, Mass.: Belknap Press, Harvard University, 1982).

2. Two of these shrews *(Sorex cristatus, Sorex aquaticus)* have since been identified as moles rather than shrews and transferred out of *Sorex*.

3. See Chapter 19: Mammoth Tales.

4. See Chapter 5: Aromatic Man.

5. See Chapter 12: Bonehunter's Stew.

6. See Chapter 17: Designed by a Camel.

## CHAPTER 3: Lesser-Known Principles

1. See Chapter 2: The Naming of the Shrew.

2. L. Van Valen, "A New Evolutionary Law," *Evolutionary Theory*, 1, pp. 1–30.

3. R. Lewin, "Red Queen Runs into Trouble?," *Science*, 227, 1985, p. 399.

## CHAPTER 4: Begging the Question

1. D. F. Glut, *The Dinosaur Dictionary* (Secaucus, N.J.: Citadel Press, 1972); A. Charig, *A New Look at the Dinosaurs* (New York: Mayflower Books, 1979).

2. Robert T. Bakker, *The Dinosaur Heresies* (New York: William Morrow, 1986).

3. L. B. Halstead and G. Caselli (illustrator), *The Evolution and Ecology of the Dinosaurs* (Eurobook 1975), p. 61.

4. Sandy Fritz, "Tyrannosaurus Sex: A Love Tail," *Omni*, 10(5), pp. 64–68, 78.

5. See W. J. Holland, *To the River Plate and Back* (New York: G. P. Putnam's Sons, 1913).

6. See Chapter 24: Body Double: Duplicating Dinosaurs.

## CHAPTER 5: Aromatic Man

1. See S. J. Gould, *The Panda's Thumb* (New York: W. W. Norton, 1980).

2. J. H. Winslow and A. Meyer, "The Perpetrator at Piltdown," *Science 83*, 4(7), September 1983, pp. 32–43.

## CHAPTER 6: Art on the Rocks

1. Louis S. B. Leakey, *Adam's Ancestor: The Evolution of Man and His Culture*, 4th ed. (New York: Harper & Row, 1960).

2. Richard E. Leakey, *The Making of Mankind* (New York: E. P. Dutton, 1981).

## CHAPTER 7: All About Eve

1. *Larousse World Mythology*, ed. P. Grimal (New York: Putnam, 1965).

2. R. L. Cann, M. Stoneking, and A. C. Wilson, "Mitochondrial DNA and Human Evolution," *Nature* 325, 1987, p. 31.

3. *Newsweek*, January 11, 1988, pp. 46–52.

4. A. Latorre, A. Moya, and F. J. Ayala, "Evolution of Mitochondrial DNA in *Drosophila subobscura*," *Proceedings of the National Academy of Sciences*, 83, 1986, pp. 8649–8653.

5. C. Patterson, ed., *Molecules or Morphology in Evolution: Conflict or Compromise.* (New York: Cambridge University Press, 1987).

6. See Chapter 17: Designed by a Camel.

7. See Chapter 25: Scientific Literacy.

## CHAPTER 8: The *Archaeopteryx* "Hoax"

1. See *The Beginnings of Birds* (Eichstatt: Freunde des Jura-Museums, 1985).

2. See Chapter 5: Aromatic Man.

3. *Creation Research Society Quarterly*, September 1983, pp. 121–122.

4. See Chapter 22: Monkey Trials.

5. *British Journal of Photography*, 132, pp. 264–266; 348; 358–359, 367; 468–470; 693–695, 703 (1985).

6. See Chapter 24: Body Double: Duplicating Dinosaurs.

7. Gavin de Beer, *Archaeopteryx lithographica.* British Museum of Natural History, Publication 224, 1954.

8. A. J. Charig, F. Greenaway, A. C. Milner, C. A. Walker, and P. J. Whybrow, "*Archaeopteryx* Is Not a Forgery," *Science*, 232, 1986, pp. 622–626.

## CHAPTER 10: Gregorian Chance

1. Or 365.24220 days. The *sidereal year* of 365 days 6 hours 9 minutes 9.5 seconds (365.25636 days) measures the revolution

of the earth around the sun from a given star back to the same star. The *anomalistic year* of 365 days 6 hours 13 minutes 53 seconds (365.25964 days) is the period during which the earth passes from one point in its orbit (e.g., perihelion) back to the same point.

## CHAPTER 11: Backing into the Future

1. R. Lindsey, "Teachings of 'Ramtha' Draw Hundreds West," *New York Times*, November 16, 1986, pp. 1, 14.

2. See Chapter 23: Natural Predictions.

3. R. L. Wesson and R. E. Wallace, "Predicting the Next Great Earthquake in California," *Scientific American*, 242(2), 1985, pp. 35–43.

4. R. A. Kerr, "Suspect Terranes and Continental Growth," *Science*, 222, 1983, pp. 36–38; John McPhee, *In Suspect Terrain* (New York: Farrar, Straus & Giroux, 1983).

## CHAPTER 13: Local Heroes

1. See Chapter 12: Bonehunter's Stew.

2. Hatcher's life and contributions to paleontology are elegantly summarized by G. G. Simpson in *Discoverers of the Lost World* (New Haven: Yale University Press, 1984), a history of paleontological explorations in South America.

3. See *Bone Hunters in Patagonia: Narrative of the Expedition* (Woodbridge, Conn.: Ox Bow Press, 1985), Hatcher's detailed account of the three Patagonian expeditions.

4. E. H. Colbert, *Men and Dinosaurs* (New York: E. P. Dutton, 1968).

## CHAPTER 15: 'Til the Cows Come Home

1. Juliet Clutton-Brock, *Domesticated Animals from Early Times* (Austin: University of Texas Press, 1981).

2. Alfred W. Crosby, Jr., *The Columbian Exchange: Biological and Cultural Consequences of 1492* (Westport, Conn.: Greenwood Press, 1973).

3. F. E. Zeuner, *A History of Domesticated Animals* (New York: Harper & Row, 1963).

## Chapter 16: Oat Cuisine

1. Joe Ben Wheat, "The Olsen-Chubbuck Site: a paleo-Indian bison kill," *American Antiquity*, 37(1), 1972, pp. 1–180.

2. See Chapter 3: Lesser-Known Principles.

3. Bruce N. Ames, "Dietary Carcinogens and Anticarcinogens," *Science*, 221, 1983, pp. 1256–1264.

## Chapter 19: Mammoth Tales

1. *Transactions of the American Philosophical Society*, Volume 23, Part 1, 1929.

2. See Chapter 6: Art on the Rocks.

3. P. S. Martin and R. G. Klein, eds., *Quaternary Extinctions: A Prehistoric Revolution* (Tucson: University of Arizona Press, 1984); also, a review of this volume: L. Krishtalka, "The Pleistocene Ways of Death," *Nature*, 312, 1984, p. 225.

4. B. Bower, "Extinctions on Ice," *Science News*, 132, 1987, p. 284.

5. N. Owen-Smith, "Pleistocene Extinctions: The Pivotal Role of Megaherbivores," *Paleobiology*, 13, 1987, p. 351; R. Lewin, "Domino Effect Invoked in Ice Age Extinctions," *Science*, 238, 1987, p. 1509.

6. *Science News*, October 3, 1987, p. 215.

7. B. Kurtén and E. Anderson, *Pleistocene Mammals of North America* (New York: Columbia University Press, 1980).

## CHAPTER 20: Chewing the Cud

1. Richard Owen, *On the Anatomy of Vertebrates, Vol. 3: Mammalia* (Longman's Green and Co., 1868).

2. Reay Tannahill, *Food in History* (New York: Stein and Day, 1973).

3. Marvin Harris, *Cows, Pigs, Wars and Witches* (New York: Vintage Books, 1978); *Cultural Materialism* (New York: Random House, 1979).

## CHAPTER 21: Face Value

1. *La Physionomie Humaine: Son Mechanisme et Son Role Social* (Paris: Alcan, 1907).

2. R. B. Zajonc, "Emotion and Facial Efference: A Theory Reclaimed," *Science*, 228, 1985, pp. 15–21.

3. A. J. Fridlund and A. N. Gilbert; C. E. Izard; A. N. Burdett; R. B. Zajonc, "Emotions and Facial Expression" (letters), *Science*, 230, 1985, pp. 607–610, 687.

4. G. Hewes, *Current Anthropology*, 14, February/April 1973, pp. 5–24.

5. B. G. Campbell, *Humankind Emerging*, 2nd ed. (Boston: Little, Brown & Co., 1979).

## CHAPTER 22: Monkey Trials

1. *Edwin W. Edwards, etc., et al., appellants* v. *Don Aguillard et al.* on appeal from the United States Court of Appeals for the Fifth Circuit, Supreme Court of the United States No. 85-1513.

## CHAPTER 23: Natural Predictions

1. L. Krishtalka, "Missing Links: Natural Predictions," *Carnegie Magazine*, January/February 1983.

2. L. Krishtalka, "Missing Links: Mid-Season Batting Average," *Carnegie Magazine*, July/August 1983.

3. See Chapter 11: Backing into the Future; also *Time*, December 7, 1987, p. 62.

4. See *Newsweek*, April 11, 1983, p. 74.

5. John Reader, *Missing Links* (Boston: Little, Brown & Co., 1981); Richard Leakey, *The Making of Mankind* (New York: E. P. Dutton, 1981); Donald Johanson and Maitland Edey, *Lucy: The Beginnings of Humankind* (New York: Simon and Schuster, 1981).

6. L. Krishtalka, "What's Bred in the Bone," review of *Laetoli: A Pliocene Site in Northern Tanzania*, ed. M. D. Leakey and J. M. Harris (London: Clarendon Press, 1987), *Nature*, 329, 1987, pp. 679–680.

7. Heinz R. Pagels, *The Cosmic Code* (New York: Simon and Schuster, 1982).

## CHAPTER 24: Body Double: Duplicating Dinosaurs

1. E. P. White, "On the Duplication of Dinosaurs," *Rubber Developments*, Vol. 40, No. 2, 1987.

2. See Chapter 4: Begging the Question.

3. *Annals of Carnegie Museum*, Vol. 3, No. 3, 1905, p. 447.

4. F. T. Stewart and C. P. Stewart III (New York: Harper and Row, 1988).

5. *Weekly World News*, April 12, 1988.

6. "Dinosaur Park in Texas Is Struggling to Survive," (*New York Times*, November 30, 1986.

7. Lynn Barber, *The Heyday of Natural History: 1820–1870* (London: Jonathan Cape, 1980). E. H. Colbert and K. Beneker, "The Paleozoic Museum in Central Park, or the Museum That Never Was," *Curator*, 2(2), 1959, pp. 137–150.

## CHAPTER 25: Scientific Literacy

1. Published by Houghton Mifflin (New York, 1987).

2. *The New York Times Book Review*, June 7, 1987, p. 5.

3. See Chapter 22: Monkey Trials.

4. *Earth Science*, 2nd ed. (Columbus: Charles E. Merrill Publishing Co., 1979).

5. See *Rising from the Plains*, by John McPhee (New York: Farrar, Straus & Giroux, 1986), a superb work about Rocky Mountain geology, Dave Love and the Old West in Wyoming.

6. See Chapter 1: Dinosaur Plots.

7. See Chapter 17: Designed by a Camel.

# Index

## D

## F

## G



## H

## N

## U



## V

## W



## X



## Y



## Z